Contents

1

Introduction

One of the main aims of this book is to explore in greater depth and detail one of the Integrating Themes running through *Environment and People*: **Natural Hazards (IT2).** For all the advances that society has made during the 20th century, we are still learning to live with natural hazards. Our awareness of them has not been diminished in any way by advances in scientific knowledge, engineering and technology. Bearing in mind the huge investments that people have made in buildings and transport in many parts of the world, the need to be vigilant about the risks of hazards is perhaps greater than it has ever been.

SECTION A

Definitions

1 Atmospheric		2 Hydrological
Single element	*Compound hazards*	Flooding: riverine (rain,
Rain	Rain and wind storms	snowmelt, natural damburst
Freezing rain ('glaze')	'Glaze' storms	floods)
Hail	Thunderstorms	Lake and sea-shore wave action
Snow	Tornadic storms and tornadoes	Waterlogging
Wind	Hurricanes	Sea-ice and icebergs
Lightning	Blizzards	Runoff
Temperature: 'heat wave,'	'White-out'	Drought
'cold spell', frost	Drought	Glacier advance
Fog		
3 Geological	**4 Biological**	**5 Technological**
Mass-movements: landslides,	Severe epidemics in humans	Transport accidents
avalanches, mudflows, subsidence, etc.	Severe epidemics in plants	Industrial explosions and fires
Erosion (foundations, soils, etc.)	Severe epidemics in domestic	Accidental releases of toxic gas
Silting (dykes, rivers, harbours, farmland)	and wild animals	Nuclear power plant failures
Earthquakes	Animal and plant invasions	Failures of public building
Volcanic eruptions	(e.g. locusts)	or other structures
Shifting sands	Forest and grassland fires	Germ or nuclear warfare

Figure 1.1 The main types of hazard

A **hazard** may be defined as an event or process that threatens, or actually causes damage and destruction to, people and their settlements. A **natural hazard** is one produced by environmental processes and involves events such as storms (atmospheric), flooding (hydrological), earthquakes and volcanic eruptions (geological). A possible criticism of the hazard classification given in **1.1** is that it includes geomorphological hazards, such as landslides, coastal erosion and silting, under the geological heading. It needs to be stressed at the outset that a natural event only becomes a natural **hazard** when it causes damage to property and loss of life. There are two other types of hazard. **Biological hazards** are created by the spread of diseases and pests and when they assume epidemic proportions. As with fire, they can have a devastating effect not just on people but also on plant and animal life. Finally, there are those hazards that are **created by people**. These range from air crashes to the collapse of buildings, from industrial explosions to germ and nuclear warfare. Essentially, they are hazards related to technology; many of them are of an accidental nature.

To cover all the hazards listed in **1.1** in a book of this size would be an impossible task. For this reason, this book focuses on a selection of natural hazards falling into the geological, geomorphological and atmospheric categories; namely earthquakes, volcanoes, landslides, severe weather, and flooding. Biological hazards, including drought, will be covered in another book in the EPICS series.

There are some other terms that need to clarified early on, because they are directly related to natural hazards. The first is **risk**. Risk is all about the probability of a particular hazard occurring and adversely affecting people. Risk is also what people take knowing that they are exposed to a natural event that might prove hazardous. The greater the probability that a natural event will occur in a particular area and prove hazardous, the greater the risk people run by either remaining in that area or failing to take appropriate evasive action (referred to as **adjustment** or **mitigation**). Another term is **disaster**. This relates to the actual hazardousness of a natural event. If a strong wind causes a few chimney-pots to be dislodged, then this scarcely qualifies as a serious hazard. If, however, as in the great hurricane that hit southern England in 1982, there are fatalities and extensive damage to property, then the natural event not only becomes a serious hazard, but it is raised to the status of a disaster. Figure **1.2** shows

Figure 1.2 The global incidence of five main disaster types based on the number of events and their impacts, 1964–89

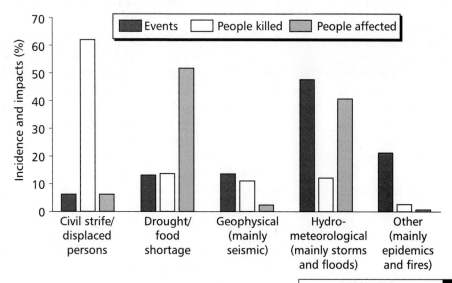

the five main causes of disasters worldwide. The graphs show the frequency of events and their impacts. Notice that civil strife (essentially a hazard created by people) leads to more deaths than the natural hazards. Perhaps the hazard classification in **1.1** should have added to it at least one more category to include such events as wars, revolutions and ethnic cleansing.

Review

1 Define the following terms:
- hazard
- natural hazard
- risk
- disaster.

2 Refer to the information in **1.1**.
  a Describe the variety of natural hazards that exist.
  b Which natural hazards are experienced in your home country?
  c Which of those hazards do you think are the most serious?

3 Study **1.2**.
  a Which type of disaster:
  - occurs most frequently
  - is responsible for most fatalities
  - affects the greatest number of people?
  b Which type of disaster do you think is the greatest hazard? Justify your answer.

4 Suggest some measures for establishing when a natural hazard becomes a disaster.

Are natural hazards entirely 'natural'?

Whilst events such as landslides, earthquakes and hurricanes are undoubtedly 'natural' events involving natural processes, there is an increasing human influence and involvement both in terms of their causes and their effects.

Earthquakes
- Human activities such as coal mining, reservoir construction and water abstraction can trigger earthquakes.
- Poor-quality buildings in earthquake zones may lead to many deaths and injuries.

Volcanoes
- People may choose or, if land is in short supply, be forced, to live on the flanks of active volcanoes. Volcanic soils are often fertile and so encourage farming and settlement.

Landslides
- Many human factors can cause a slope to become unstable, e.g. building, excavation, altering slope drainage, overgrazing or deforestation.
- Potentially unstable slopes are often built upon, particularly where land is in short supply, for example in rapidly expanding cities like Lima in Peru.

Atmospheric hazards

- Many people choose to live in or develop coastal areas that are known to be at risk from hurricanes.
- People often venture out into storms, for example blizzards, completely unprepared for the dangers.

Floods

- Floodplains are often very fertile, and so attract farming and settlement almost regardless of the risks.
- Deforestation of slopes increases runoff and so often leads to flooding.

We have already established that if there were no people, there would be no such thing as a natural hazard. It is the threat to people and their property that converts a natural event into a hazard. Equally important, though, is what emerges from the above examples, namely that people can also:

- trigger potentially hazardous natural events through everyday acts such as farming, cutting fuelwood and building homes
- increase the disaster potential of hazards by taking risks with the natural environment.

In short, natural hazards and disasters are an integral part of human life. This is especially the case in the developing world where the pressures of survival are often such that they encourage high levels of risk taking.

Review

5 In what ways can people induce natural hazards?

6 How can the actions (or the inactions) of people cause a natural event to become a natural disaster?

SECTION C

Living with natural hazards – a risky business

Natural hazards are widely reported in the media. Earthquakes, tornadoes and hurricanes make great stories. It is, however, important to get a sense of perspective and to try to understand the risks involved.

- Volcanic eruptions are not common events in a single lifetime, and volcanoes often give warning signs of an impending eruption. Maybe it is not so unreasonable to live and farm on the flanks of a volcano, especially if there are no suitable alternatives?
- Earthquakes are a greater risk, as they occur unannounced in many of the world's most heavily populated areas. Living in an earthquake zone is very risky, but the risk of death, injury and damage to property can be reduced through public awareness and by strict planning and building controls.
- Landslides are most common in very remote, mountainous environments where few people would choose to live anyway. However, in urban areas, there are risks associated with the landslide hazard. For example, in Hong Kong, the lack of flat land means that many developments have had to be built on steep slopes. The weak soils and heavy rainstorms can readily trigger the collapse of these slopes. The risks can be reduced by careful

planning and making use of engineering solutions aimed at slope stabilisation.

- The risks associated with storms vary enormously. Tropical cyclones (hurricanes), for example, cause most damage on low-lying coasts. The closer to the sea that one chooses to live, the greater the risks. Further inland, though the hazard (a storm surge) remains the same, the risk is considerably less.
- Flooding is one of the most widespread and frequent of the natural events (**1.2**). The risk of such events becoming hazards and disasters is high for the simple reason that human settlement is drawn to rivers and their floodplains. They provide water, a medium of transport and readily cultivable fertile land. The risks of flooding can certainly be reduced by building up the banks of rivers and introducing measures that control peak discharge.

Risk is closely related to a person's or a group's perception of a hazard and about decision-making. Deciding to cross the road, taking a trip in an aeroplane or doing a bungee jump all involve some sort of risk assessment. Some people are prepared to take greater risks than others. This is often (but not always) the case with natural hazards. The brutal truth is that some people, particularly in developing countries, do not have much choice when it comes to evading natural hazards.

People may choose to live close to the sea for all the obvious advantages that such a location offers, despite the fact that a hurricane 'might' hit them at some time in the future. In the USA some people follow tornadoes as a hobby. These 'stormchasers', as they are known, often take considerable risks to get close to a tornado and to take photographs. Many Californians just accept the risk of an earthquake, enjoying life to the full and not worrying too much about what 'might' happen.

Having established the risk, the next stage is to map out a number of different strategies or courses of action. These may range from deciding to do nothing, to different forms and degrees of what is called **adjustment** or **mitigation**. Adjustment is taking actions that seek to reduce, minimise or eliminate the threat posed by a natural event.

Those adjustment options will be costed out and eventually a choice will be made (it could be to do nothing at all). The decision-making process will involve balancing the costs of adjustment against the costs of any damage that might be sustained if nothing is done. The decision-making will also be influenced by the resources (financial, technological, etc.) that are available. The fewer those resources, the more likely that adjustment will be rather minimal. The accuracy of the hazard perception and risk assessment will also be a key factor in the choice of adjustment.

As you read about the different natural hazards in this book, try to assess:

- how 'natural' they really are
- to what extent and in what ways people have triggered them
- how people have turned them into disasters
- how far people have undertaken risk assessments when making decisions about possible adjustments.

Review

7 a Suggest three different adjustments that might be made to reduce the risks of coastal erosion along a populated stretch of British coastline.
  b Which of your three adjustments do you think would be (i) most expensive and (ii) most feasible? Justify your answers.

8 'Risk assessment is conditioned by hazard perception.' Discuss and illustrate this statement.

2

Earthquakes

Introduction

An earthquake is a sudden and brief period of intense ground-shaking. Movement is often in a horizontal as well as a vertical plane, and people caught in an earthquake frequently refer to a 'rolling' motion akin to being on a rough sea. The following is an eyewitness account of the Alaskan earthquake of 1964. The earthquake, at 9.2 on the Richter scale, was the second most severe earthquake ever recorded (the most severe occurred in Chile in 1960 and measured 9.5). Betty Feakes was baby-sitting in the small port of Valdez (most of which was destroyed by the earthquake):

> The three kids and I had just finished dinner when I heard a thump and a rumbling coming. I thought it was the snow plow [plough] out in the front, and had just walked to the arch in the living room when something hit the house with a terrible jolt, and then everything came apart at the seams.
>
> The house started to rock and roll. Things started falling off the walls and the shelves. The two kids in the living room were knocked to the floor and started screaming. I looked for the baby still in his high chair and saw his chair rocking across the floor heading for the stove. The floor was rolling and heaving like a live thing in torment.

Being caught in an earthquake is a terrifying experience because, although it may only last for tens of seconds, the damage it inflicts can be immense and, of course, during the earthquake itself, the fear of the unknown can be overpowering. Even if there is no injury or damage to property, an earthquake victim is likely to be in shock for some time after the event.

Earthquakes are measured using an instrument called a **seismograph**. The trace produced by a seismograph is called a **seismogram**. Earthquake magnitude (strength) is measured according to the **Richter scale**. This is a logarithmic scale which has no upper limit. Being logarithmic, there is a ten-fold increase in magnitude between each point on the scale. The most damaging earthquakes have a value of between 5 and 7 on the Richter scale. Figure **2.1** lists the world's most destructive earthquakes of all time. Notice that the magnitude equals or exceeds 7.5 in all cases. The physical effects of an earthquake are measured using the **Modified Mercalli scale** (**2.2**, page 12).

Date	Location	Deaths (est.)	Magnitude	Comments
23 January 1556	China, Shansi	830 000		
27 July 1976	China, Tangshan	255 000	8.0	
9 August 1138	Syria, Aleppo	230 000		
22 December 856	Iran, Damghan	200 000		
16 December 1920	China, Gansu	200 000	8.6	Major fractures, landslides
22 May 1927	China, near Xining	200 000	8.3	Large fractures
23 March 893	Iran, Ardabil	150 000		
1 September 1923	Japan, Kwanto	143 000	8.3	Great Tokyo fire
September 1290	China, Chihli	100 000		
28 December 1908	Italy, Messina	70 000–100,000	7.5	Deaths from earthquake and tsunami
November 1667	Caucasia, Shemakha	80 000		
18 November 1727	Iran, Tabriz	77 000		
1 November 1755	Portugal, Lisbon	70 000	8.7	Great tsunami
25 December 1932	China, Gansu	70 000	7.6	
31 May 1970	Peru	66 000	7.8	US$530 000 damage, great rock slide, floods
1268	Asia Minor, Silicia	60 000		
11 January 1693	Italy, Sicily	60 000		
30 May 1935	Pakistan, Quetta	30 000–60 000	7.5	Quetta almost completely destroyed
4 February 1783	Italy, Calabria	50 000		
20 June 1990	Iran	50 000	7.7	Landslides

Figure 2.1 The most destructive known earthquakes

Source: USGS National Earthquake Information Center

Review

1 a What is the difference between the Richter scale (**2.1**) and the Modified Mercalli scale (**2.2**)?
b Why is it somewhat questionable to link the two together?

I	Not felt except by a very few under especially favourable circumstances.
II	Felt only by a few people at rest, especially on upper floors of buildings.
III	Felt quite noticeably indoors, especially on upper floors of buildings, but many people do not recognise it as an earthquake. Standing vehicles may rock slightly. Vibration like passing truck. Duration estimated.
IV	During the day, felt indoors by many, outdoors by a few. At night some awakened. Dishes, windows, doors disturbed; walls make cracking sound.
V	Felt by nearly everyone, many awakened. Some dishes, windows, etc. broken; a few instances of cracked plaster; unstable objects overturned. Pendulum clocks may stop.
VI	Felt by all, many frightened and run outdoors. Some heavy furniture moved; a few instances of fallen plaster or damaged chimneys. Damage slight.
VII	Damage negligible in buildings of good design and construction; slight to moderate in well-built ordinary structures; considerable in poorly built or badly designed structures; some chimneys broken.
VIII	Damage slight in specially-designed structures; considerable in ordinary substantial buildings, with partial collapse; great in poorly-built structures. Fall of chimneys, factory stacks, columns, monuments, walls. Heavy furniture overturned.
XI	Damage considerable in specially-designed structures; well-designed frame structures thrown out of plumb; great in substantial buildings, with partial collapse. Buildings shifted off foundations.
X	Some well-built wooden structures destroyed; most masonry and frame structures destroyed with foundations; ground badly cracked. Rails bent.
XI	Few, if any (masonry) structures remain standing. Bridges destroyed. Rails bent greatly.
XII	Damage total. Practically all works of construction are much damaged, or destroyed.

Figure 2.2 Modified Mercalli earthquake intensity scale

SECTION B

What causes an earthquake?

An earthquake commonly results from the sudden release of energy deep underground. This release of energy usually takes the form of a jerk or slippage along a line of weakness called a fault. Shock waves called **seismic waves** radiate out in all directions from the point of slippage – the **focus** (*Environment and People*, figure **2.10** page 17). It is these shock waves that cause the shaking on the ground surface. Generally, although by no means exclusively, the greatest effect is felt nearest the **epicentre** – the point on the surface immediately above the focus of the earthquake.

Most of the world's active faults lie at the edges of the plates, which is why the majority of earthquakes occur at plate margins (see *Environment and People*, figure **2.4** page 12). It is not hard to appreciate how, with the tremendous forces at work at plate boundaries, particularly at destructive and conservative boundaries, regular slippages result. However, if you were to plot the incidence of earthquakes on a map showing the plate margins, you would notice that some do not occur at those margins.

Look at **2.3** showing the distribution of earthquake epicentres in the North Atlantic region. Notice the neat line marking the mid-Atlantic ridge (a constructive margin), but also compare this with the chaos in Western Europe. Even the UK has occasional earthquakes. In 1990 an earthquake of over 5 on the Richter scale struck Clun in Shropshire, and tremors were felt as far away as Exeter, London and Newcastle, although no significant damage was done.

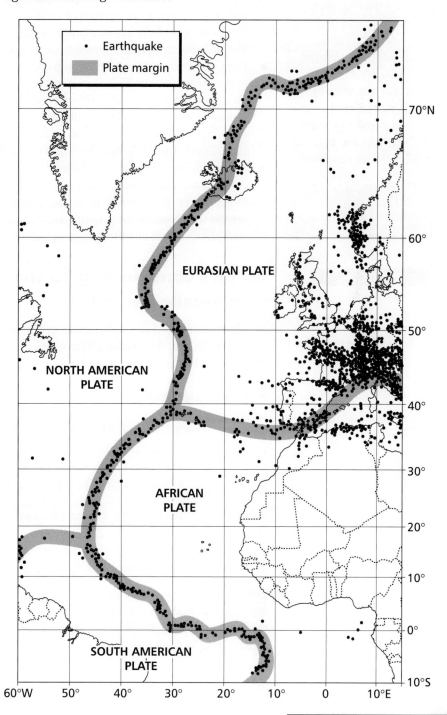

Figure 2.3 Earthquakes in the North Atlantic region, 1971–86

There are a number of possible causes of these mid-plate earthquakes.

Review

2 Study **2.3**.
 a Explain the origin of earthquakes along the mid-Atlantic ridge.
 b Suggest possible causes of earthquakes in:
 ▪ the coal-mining areas of the UK, Belgium and Germany
 ▪ the North Sea basin.

▪ Underground subsidence following deep mining. This is a known cause of earthquakes in the UK coalfield areas, for example around Stoke-on-Trent, although it is less significant now that the industry has declined.
▪ Lubrication of an ancient fault by migrating water or oil. Whether the liquids actually lubricate the fault (in the way that oil lubricates the moving parts of a car engine) or simply weaken the rock either side of the fault is not fully known, but there is substantial evidence that it does lead in some way to increased seismic activity.
▪ Dam and reservoir construction have been linked to increased earthquake activity. This is partly due to the increased pressure exerted on a fault from above and also the likelihood of water seeping into the groundwater zone and lubricating a fault. In Zambia, hundreds of earthquakes a year occurred as Lake Kariba filled up between 1958 and 1963. Soon after 1963 the earthquakes stopped.
▪ Scientists now believe that the extreme pressures exerted at plate margins actually cause a web of cracks (faults) to form right across the plate, rather like a windscreen that crazes when hit by a stone. This means that it would be possible for marginal tensions to be released mid-plate as well as at the plate margin itself (see page 17).

SECTION C

Factors contributing to the earthquake hazard

The Earth is rocked by earthquakes every day, yet not every earthquake causes death and destruction. There are several factors that can turn an earthquake into a natural disaster.

Earthquake magnitude

In general, the more powerful the earthquake the more potential damage it will do. Figure **2.1** on page 11 shows this to be broadly correct. However, the San Francisco earthquake of 1989 registered 7.1 on the Richter scale but only 65 people were killed, whereas a less severe earthquake measuring 6.8 killed over 25 000 people in Armenia a year earlier.

Ability of buildings to withstand shaking

Very often, the huge death toll resulting from an earthquake is a consequence of poor building design rather than a strong earthquake. There is a saying: 'Earthquakes don't kill people, buildings do!'

Poor building design was particularly evident in Mexico City in 1985. Both low-rise and high-rise buildings collapsed as walls gave way and ceilings crashed down onto the floors below. Some buildings toppled over and others literally sank into the ground.

Population distribution

Nearly half of the world's largest cities are located in earthquake-prone areas, particularly around the Pacific Ocean. One of the most vulnerable cities is Tokyo. In 1923, 143 000 people lost their lives and two-thirds of the city was flattened. The consequences of a similar earthquake happening today in a city with over 12 million people would be enormous.

Nature of the underlying geology

Different materials respond in different ways to earthquakes. Sands and clays may become liquefied (jelly-like), so promoting landslips and building collapse. This was a significant factor in the Mexico City earthquake of 1985. Much of the city lies on old lake-bed sediments. In the 1989 San Francisco earthquake, the most severe damage to property was in the Marina District. The housing here had been constructed on landfill material which was basically unstable.

Wealth of a country

In theory, the wealthier a country, the better able it is to build appropriately and respond effectively to the earthquake hazard. Certainly, building standards are lower in some of the poorest countries of the world, and authorities, having less money and emergency resources, are far less able to cope with the aftermath of an earthquake.

Review

3 Describe some of the ways in which poor building design can contribute to the high death toll often experienced in urban areas.

SECTION D

What are the effects of an earthquake?

Earthquakes can have devastating and far-reaching effects. Most often we associate the earthquake hazard with ground-shaking and the subsequent collapse of buildings. However, the effects can be much more long-term.

It is helpful to identify both short-term and longer-term effects of earthquakes.

Short-term effects

- People may be killed or injured.
- Buildings collapse, trapping or burying their inhabitants (see **2.4** on page 17).
- Bridges collapse, roads break up and railways become like roller-coaster track.
- Communications break down.
- Water, gas and sewage pipes rupture.
- Leaking gas ignites, causing fires to break out. This may well lead to more deaths than building collapse.
- In mountainous areas earthquakes can trigger landslides.
- If an earthquake is centred out at sea, the intense shaking can lead to the creation of huge waves called **tsunami**. These waves (often

inaccurately referred to as 'tidal waves') may be several metres in height and can inundate coastal areas well away from the earthquake epicentre. They have been responsible for some of the greatest death tolls associated with earthquakes.

■ Tremendous worry and anxiety amongst people caught in the earthquake.

Longer-term effects

■ Homelessness and lack of shelter. This is a particular problem in developing countries and in all countries in the winter.
■ If communication lines are broken it might take several days or weeks for help to arrive.
■ Disrupted services may lead to disease as there will be little fresh water and sewage disposal.
■ Families may be torn apart by an earthquake. Large numbers of orphans may be created and those involved in the earthquake may have long-term emotional problems.
■ The cost of re-building will often be immense and time-consuming. The poorer countries of the world may not have sufficient resources.
■ Without homes and jobs, people will migrate, so becoming refugees, unable to fend for themselves.

Review

4 a The longer-term effects of an earthquake are often overlooked by the media. Why do you think this is so?
  b Compare the likely longer-term effects of an earthquake in a developed country with those in a developing country.
  c Why are the longer-term effects of an earthquake often more far-reaching, particularly in developing countries, than the short-term effects?

SECTION E

Case study: The Khillari earthquake (India), 1993

At 3.45am on 30 September 1993 a powerful and totally unexpected earthquake hit western India. The earthquake, measuring 6.4 on the Richter scale, was centred near Khillari some 400km to the south-east of Bombay. More than 25 000 people were killed and 50 villages and towns were destroyed. Figure **2.4** is an extract from *The Daily Telegraph* describing the immediate effects of the earthquake.

Several factors contributed to the disaster:

INDIAN QUAKE KILLS 16 000 AS THEY SLEEP

By Rahul Bedi in New Delhi

Sixteen thousand people died when an earthquake destroyed their homes as they slept in a large area of western India yesterday. Television reports said at least another 12 000 people were injured as two towns and about 30 villages were razed in Maharashtra state south-east of Bombay.

The 10-second quake, registering between 6.0 and 6.4 on the Richter scale and said to be the most devastating in the subcontinent since 1935, occurred at about 4am, bringing roofs crashing down. Witnesses said huge cracks opened in the ground, swallowing whole houses.

'The rising sun created darkness for us this morning, swallowed up our villages, and made our houses into tombs,' a survivor told one of the first reporters to reach the scene.

The districts of Latur and Osmanabad bore the brunt of the earthquake. Officials said most of Khillari, a sugar cane and grape growing township in Latur district with a population of 180 000, and the town of Umarga, were destroyed. The largest death toll – 3050 – was reported from Umarga, where 60 per cent of the buildings were razed, but 80 per cent of Khillari was said to

have been flattened with at least 1000 people dead and hundreds trapped.

'So many people have died that we just cannot count,' said one official.

'Those few seconds seemed to last forever,' said Solani Bhagwat, 35. 'I didn't know how it happened. It was dark and I could hear people shrieking and howling. Only when the sun came out did I realise they were all trapped in their houses.'

Inside Khillari's small, overcrowded hospital, hundreds of injured people, many bleeding and wailing, lay on the floors or in a tent set up in a courtyard. 'We don't have enough drugs and bandages,' said Dr Ahiwin Solekar.

About 50 villages were said to have suffered severe damage. 'The death toll is going up by the minute,' said Praveen Pardesi, a senior government administrator.

As police and civilians freed buried survivors, screaming and writhing in pain, their continuing rescue attempts were often hampered by stunned and homeless people who crowded around the debris or roamed through the destroyed villages praying for their lost relatives.

Soldiers and policemen from across India rushed to the remote area, bringing stretchers, tents, medical supplies, earth movers, bulldozers and mobile hospitals. But relief workers had trouble reaching some villages that had recently lost roads and bridges to heavy monsoon rains.

Several areas could not be reached because approach roads had been ruptured by the quake, which also severed electricity supplies, plunging the entire region into darkness.

Figure 2.4 The effects of the Khillari earthquake

Source: *The Daily Telegraph*, 1 October 1993

- The focus of the earthquake was very shallow. This meant that the shock waves had a more destructive effect.
- The earthquake happened in the middle of the night when most people were in their homes.
- The region was completely unprepared for earthquakes.
- Building standards were low.
- The area had a high and rapidly growing population. Some 3 million people lived in the affected area.
- India is a poor country and many of its people are very poor, particularly in rural areas.

The cause of the earthquake has aroused considerable debate because no plate margin runs through central India. No wonder the earthquake was unexpected! Two particular suggestions have been made to explain the location of the earthquake.

Fault webs There are very considerable pressures in the Himalayas region as India crumples relentlessly into Europe. Whilst most of these pressures are released in the Himalayas (over 40 000 people have been killed by earthquakes in northern India and Nepal in the 20th century alone), it is possible that pressure release happens elsewhere along old fault lines many kilometres from the actual source of the pressure. Tensions at plate margins may cause a series of cracks or faults to run right across the plate itself. Some scientists believe that a complex web of faults lies across India and that localised movement along one of these caused the earthquake (**2.5**).

5 Read **2.4**.
  a Make a list of the immediate effects of the earthquake as reported in the newspaper article, which was published the day after the earthquake occurred.
  b Describe some of the likely longer-term effects of the earthquake.
  c Comment on the value and validity of newspaper articles written immediately after an earthquake.

6 Identify some factors that contributed to the scale of the disaster. For each, give reasons why it was a contributory factor.

7 With the aid of diagrams and sketch maps, examine the ways in which the following theories might account for the earthquake. Give your opinion of each theory.
  a Nearby movement along a plate boundary.
  b Web theory.
  c Construction of nearby reservoirs.

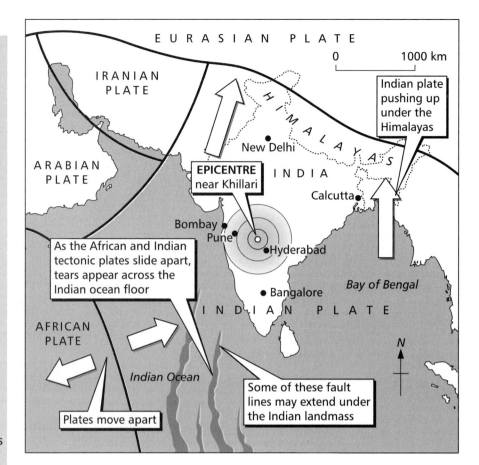

Figure 2.5 Tectonic forces at work in India

Reservoir construction Writing in the *New Scientist* (2 April 1994), Seeber suggests that the damming of the Lirna River some three years earlier may have been responsible for the earthquake. Many earthquakes in India are believed to have been triggered by dam and reservoir construction and, as most of the casualties occurred close to the dam and the river itself, it seems likely that the construction of the dam and the reservoir did indeed play some part in causing the earthquake.

Case study: The Los Angeles earthquake, USA 1994

At 4.31am on 17 January 1994, Los Angeles was hit by a magnitude 6.6 earthquake. Although some US$7 billion worth of damage was done by the earthquake, remarkably, only 40 people lost their lives (**2.6**). This compares with the Indian earthquake which, although less powerful (6.4 on the Richter scale), killed 25 000 people. It is important to consider why two earthquakes, very similar in terms of magnitude, should have dramatically different effects in terms of loss of life. Was it anything to do with the unusual nature of the earth movements (**2.7**)?

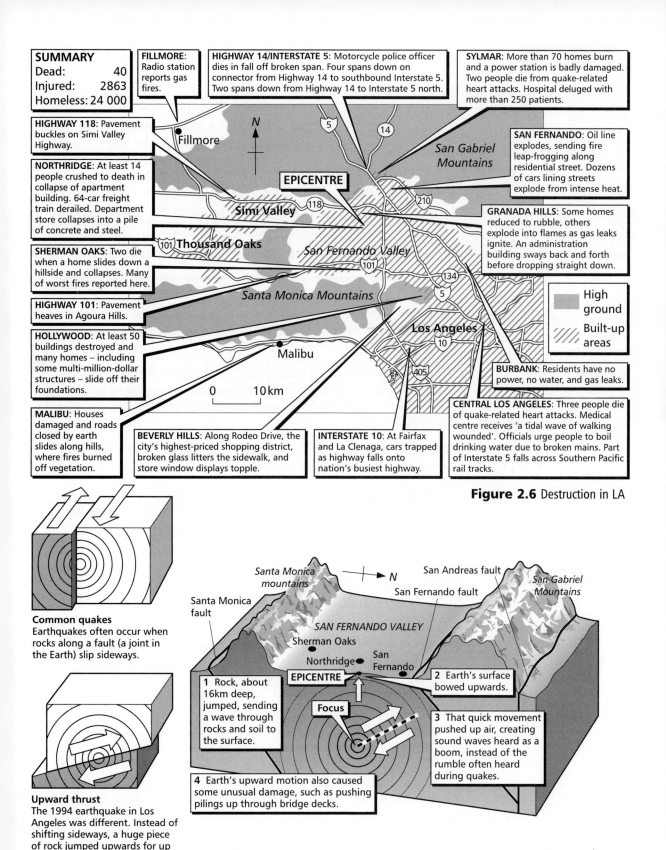

SUMMARY
Dead: 40
Injured: 2863
Homeless: 24 000

FILLMORE: Radio station reports gas fires.

HIGHWAY 14/INTERSTATE 5: Motorcycle police officer dies in fall off broken span. Four spans down on connector from Highway 14 to southbound Interstate 5. Two spans down from Highway 14 to Interstate 5 north.

SYLMAR: More than 70 homes burn and a power station is badly damaged. Two people die from quake-related heart attacks. Hospital deluged with more than 250 patients.

HIGHWAY 118: Pavement buckles on Simi Valley Highway.

SAN FERNANDO: Oil line explodes, sending fire leap-frogging along residential street. Dozens of cars lining streets explode from intense heat.

NORTHRIDGE: At least 14 people crushed to death in collapse of apartment building. 64-car freight train derailed. Department store collapses into a pile of concrete and steel.

GRANADA HILLS: Some homes reduced to rubble, others explode into flames as gas leaks ignite. An administration building sways back and forth before dropping straight down.

SHERMAN OAKS: Two die when a home slides down a hillside and collapses. Many of worst fires reported here.

HIGHWAY 101: Pavement heaves in Agoura Hills.

HOLLYWOOD: At least 50 buildings destroyed and many homes – including some multi-million-dollar structures – slide off their foundations.

MALIBU: Houses damaged and roads closed by earth slides along hills, where fires burned off vegetation.

BEVERLY HILLS: Along Rodeo Drive, the city's highest-priced shopping district, broken glass litters the sidewalk, and store window displays topple.

INTERSTATE 10: At Fairfax and La Clenaga, cars trapped as highway falls onto nation's busiest highway.

CENTRAL LOS ANGELES: Three people die of quake-related heart attacks. Medical centre receives 'a tidal wave of walking wounded'. Officials urge people to boil drinking water due to broken mains. Part of Interstate 5 falls across Southern Pacific rail tracks.

BURBANK: Residents have no power, no water, and gas leaks.

High ground

Built-up areas

0   10km

EPICENTRE

San Gabriel Mountains

Simi Valley

Thousand Oaks

San Fernando Valley

Santa Monica Mountains

Malibu

Los Angeles

Fillmore

Figure 2.6 Destruction in LA

Common quakes
Earthquakes often occur when rocks along a fault (a joint in the Earth) slip sideways.

Upward thrust
The 1994 earthquake in Los Angeles was different. Instead of shifting sideways, a huge piece of rock jumped upwards for up to 1 metre, resulting in damage unlike that in other quakes.

Santa Monica mountains

Santa Monica fault

San Andreas fault

San Fernando fault

San Gabriel Mountains

SAN FERNANDO VALLEY

Sherman Oaks

Northridge

San Fernando

EPICENTRE

Focus

1 Rock, about 16km deep, jumped, sending a wave through rocks and soil to the surface.

2 Earth's surface bowed upwards.

3 That quick movement pushed up air, creating sound waves heard as a boom, instead of the rumble often heard during quakes.

4 Earth's upward motion also caused some unusual damage, such as pushing pilings up through bridge decks.

Figure 2.7 Unusual rock shifts in the LA earthquake

8 Study the information in **2.6** and make a list of the effects of the earthquake.

9 Study **2.7**. Use diagrams to help explain the likely cause of the Los Angeles earthquake.

10 Identify the various factors that contributed to the disastrous effects of the earthquake.

11 The earthquake that hit Los Angeles in 1994 was very similar in magnitude to the one that shook western India in 1993. However, the effects of the two earthquakes were very different.
   a In the form of a table, make a comparison of the effects of the two earthquakes.
   b Suggest reasons for the differences you have identified in **(a)**.

WEB SEARCH

The United States Geological Survey (USGS) has some excellent material available through its National Earthquake Information Center (NEIC).

Search for lists of recent earthquakes produced by the NEIC and others. These can be printed and then the locations plotted on a world outline. Discover if the earthquakes occurred at plate margins or were mid-plate.

The Caribbean Disaster Emergency Response Agency (CDERA) at: www.cdera.org/ has a very good fact sheet.

Search for information on earthquakes in particular countries, such as Japan. You will find a huge amount in the USA, for example the Southern California Earthquake Center.

SECTION F

Reducing the earthquake hazard

Reducing the hazard of earthquakes takes two forms of adjustment:

■ attempting to predict where and when an earthquake will occur
■ attempting to reduce the damage of an earthquake by taking precautions and implementing appropriate planning measures.

Predicting earthquakes

In 1975 the Chinese successfully predicted an earthquake. It is, however, the only successful prediction that scientists have been able to make to date. Their prediction was based upon a number of precursors. There were several series of small tremors – 500 in 72 hours in early February 1975; wells began to bubble and animals began behaving strangely (pigs ate their own tails, for example!). At 2pm on 4 February 1975, 3 million people in southern Liaoning province were ordered to spend the night outdoors in straw huts and tents. At 7.36pm a magnitude 7.8 earthquake struck the area, flattening much of the town of Haicheng but, remarkably, only 500 people were killed. Had people not been warned, the death toll would probably have been tens of thousands. Unfortunately, the success was short-lived, for in the following year a devastating earthquake killed over 250 000 people at Tangshan. Although there had been a few tremors beforehand, there had been insufficient evidence to order an evacuation.

The prediction of earthquakes is one of the great scientific quests of the modern day. The theory of plate tectonics is quite helpful in predicting **where** earthquakes are likely to occur (though do not forget that the Indian earthquake of 1993 was completely unexpected as it occurred mid-plate), but scientists are no nearer being able to predict exactly **when**.

Prediction is only helpful if people can be given a reliable warning of an event that is about to happen in a few hours' time and if those people can then realistically be moved to a safe area in time. Prediction becomes a nonsense if warnings turn out to be false alarms. Furthermore, if there is nowhere safe for people to move to, would the authorities be better off saying nothing?

Underlying the search for a method of prediction is the concept of **elastic rebound**. This suggests that stress builds up at a constant rate but is only released periodically. As the build-up of pressure is constant, it might be expected that the release of pressure (an earthquake) occurs regularly. Therefore, by using historical records and geological evidence (such as faults), scientists should be able to plot recurrences of earthquakes and use this information to suggest the next likely date.

This concept has been developed into the **seismic gap theory**. If earthquakes along a particular fault line are plotted, gaps may appear where no earthquakes have occurred for some time. On the basis that pressure has been building up in these gaps but has yet to be released, it is reasonable to suggest that these areas are most likely to be hit by future earthquakes.

Much of the work on this theory has been based on the San Andreas fault in California where there is a very good historical and geological record of earthquakes. The 1989 Loma Prieta or San Francisco earthquake was, to some extent, expected as it occurred in a seismic gap (**2.8**). However, although it was expected, it had not been predicted to the hour, day or even week.

The seismic gap theory may make a good deal of sense but it cannot tell us when an event will occur, nor precisely where – again, think back to India 1993.

Precursors

Scientists have identified several precursors to earthquakes which may prove valuable in the future as more and more data is obtained. They include the following.

- **Foreshocks** that occur prior to the main earthquake. Several small earthquakes occurred in the months before the 1989 San Francisco earthquake.

Figure 2.8 The seismic gap at Loma Prieta before and after the 1989 earthquake

Source: US Geological Survey

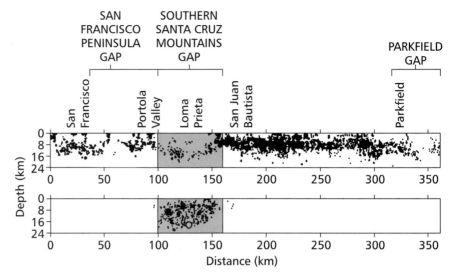

Cross-sections along the San Andreas fault from north of San Francisco to Parkfield, showing three gaps in seismicity: between San Francisco and Portola Valley, near Loma Prieta, and south-east of Parkfield. The upper section shows the location of earthquakes for the period from January 1969 to July 1989. The lower section shows the southern Santa Cruz Mountains gap after it was filled by the 17 October Loma Prieta earthquake (open circle) and its aftershocks.

- **Geophysical changes** involving localised uplift and subsidence have been identified. For example, the Niigata earthquake in Japan in 1964 was accompanied by a sudden subsidence of the coastline by up to 20cm. Research after the earthquake revealed that there had been an uplift just prior to the earthquake in much the same way that a volcano 'inflates' before an eruption.

- Changes in **electrical resistivity** may also prove to be useful in the future. The VAN method (named after its three Greek inventors) suggests that electrical currents become disturbed just prior to an earthquake and this causes a change in resistivity which can be measured. The three scientists claim to have accurately predicted 10 out of 14 earthquakes in Greece between 1987 and 1997 using this technique.

- Changes in **animal behaviour** are well documented just prior to a major earthquake, as witnessed at Haicheng in 1975. In China, panicky rats are actually an official earthquake precursor!

Planning for earthquakes

There are several measures that can be implemented in order to reduce the loss of life and keep damage to property to a minimum.

- Buildings should be designed to withstand earthquakes (**2.9a**). The separate parts (roof, walls, floors) should be securely connected and use made of cross-beams to strengthen the whole structure. Timber should be used where possible as it absorbs shock much better than bricks and mortar. Buildings should have firm and deep foundations, and high standards of design (number of storeys, gap between buildings, load-bearing columns, cables and wires in concrete, etc.)

should be rigorously enforced in areas prone to earthquakes. Heavy objects should be secured to walls, floors and ceilings so that they do not collapse on people during an earthquake, and automatic sprinkler systems should be installed to reduce the risk of fire. Automatic shutters can be used to cover window-panes, reducing the risk of broken glass showering pedestrians on the streets below. Numbers can be painted on the tops of buildings to aid emergency services after an earthquake has occurred.

- Bridges and roads should be built to withstand shaking (**2.9b**).
- People should be taught how to respond during an earthquake. In Japan and California, earthquake drills in schools are as commonplace as fire drills in the UK.
- Provision should be made for emergency relief. Food, temporary accommodation, and medical supplies should be stored outside the main earthquake zone and helicopters should be available for transport.
- Services (gas, water, electricity, etc.) should make use of flexible piping where possible, and automatic shut-offs should be used to reduce the risk of leakages.
- Earthquake hazard maps can be produced. These make use of geological information regarding soil and rock type, as well as building type and design, in order to show the areas at greatest risk from shaking, landslides, liquefaction, etc. Such maps have been drawn for many cities in Japan and the USA, but they are only any good if they are taken note of in the planning process.
- Coastal areas at risk from tsunami can be redeveloped to reduce the effects from these huge waves (see pages 31 and 75).

Figure 2.9

Earthquake-proofing:
(a) buildings and (b) bridges

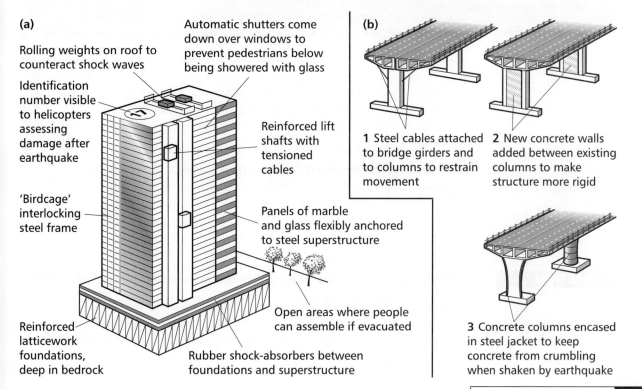

(a)

Rolling weights on roof to counteract shock waves

Identification number visible to helicopters assessing damage after earthquake

'Birdcage' interlocking steel frame

Reinforced latticework foundations, deep in bedrock

Automatic shutters come down over windows to prevent pedestrians below being showered with glass

Reinforced lift shafts with tensioned cables

Panels of marble and glass flexibly anchored to steel superstructure

Open areas where people can assemble if evacuated

Rubber shock-absorbers between foundations and superstructure

(b)

1 Steel cables attached to bridge girders and to columns to restrain movement

2 New concrete walls added between existing columns to make structure more rigid

3 Concrete columns encased in steel jacket to keep concrete from crumbling when shaken by earthquake

Of course, all these measures are fine if the country concerned is wealthy enough to be able to afford them. Many developing countries do not have the money to enforce strict regulations or to respond quickly to a disaster.

A high death toll often results from a combination of adverse circumstances. On 4 January 1998, a magnitude 6.1 earthquake struck a remote and mountainous region in northern Afghanistan, one of the poorest countries in the world. Some 4000 people were killed during and immediately after the event. The earthquake had a shallow epicentre and triggered many landslides. The remote village communities were cut off from the outside world and the poor communications meant that it was several days before relief supplies could be administered. The weather was dreadful, with snow and bitter cold certainly adding to the enormity of the disaster.

Review

12 a What is meant by 'elastic rebound'?
   b What is the seismic gap theory?
   c What evidence is there on **2.8** to suggest that the Loma Prieta earthquake of 1989 was expected?
   d According to **2.8**, which two sections of the San Andreas fault are due to have an earthquake?

13 Use the information in **2.9**, together with the text, to outline some building design measures that should be considered in earthquake-prone areas. Why do some city authorities pay scant regard to the construction of earthquake-proof buildings and bridges?

14 Should governments continue to spend money on research into earthquake prediction, or would the money be better spend preparing for the inevitable earthquake, by improving building design standards, for example?

WEB SEARCH

A good deal of research on earthquake prediction has taken place in Southern California. Search the United States Geological Survey (USGS) web sites for details. There is a site entitled 'Southern Californians cope with Earthquakes'.

Search for the Southern Californian Earthquake Center to find a great range of information advising people how to make their homes 'earthquake-proof', how to respond to an earthquake, etc. Earthquake preparedness is currently generating an enormous amount of material.

SECTION G

Extended enquiry: The Kobe earthquake, 1995

In this activity you will study some information about the Kobe earthquake that has come from a number of different sources. The aim is for you to produce a report on the earthquake examining the following aspects:

■ The effects of the earthquake.

- The causes of the earthquake.
- The factors that contributed to the scale of the disaster.
- The lessons that were learned for the future.

In preparing your report you may wish to consider the questions and suggestions in the Enquiry box on page 28. First though you should take time to read the various accounts and study the diagrams (**2.10–2.15**).

Figure 2.10 Earthquake damage in Kobe

Yasuyo Morita (17 years old)

At the time I was sleeping. First I thought my mother was waking me up, but I knew from her scream – It is an Earthquake! I didn't know what to do in the dark. The stairs are destroyed, so I went down using a ladder with bare feet wearing pyjamas. I couldn't stop my tears because of darkness and coldness. I was in a panic. There are a lot of wooden houses in my neighbourhood. Now they are destroyed without pity. Soon it got lighter and I could see my surroundings. Some went mad. I went to hospital with my grandmother who was rescued from a heap of rubble. Her finger was tearing off. We got to hospital. It was hell on earth. A man bleeding from his head, a child – purplish maybe because of suffocation. It was filled with people. My grandmother was disinfected that's all. Her injury was not serious compared with others. My grandmother's house and my grandfather's house, both are burned down. We couldn't get out anything. The town I like changed in a moment. Now I live in one of the refuges and I get scared during the night. I want to see Kobe rebuilt again.

Figure 2.11 An eyewitness account

Figure 2.12 The effects of the earthquake

On Tuesday 17 January, at 5.46am local time, an earthquake of magnitude 7.2 struck the region of Kobe and Osaka in south-central Japan. This is Japan's second-most populated and industrialised area after Tokyo, with a population of about 10 million. The shock occurred at a shallow depth on a fault running from Awaji Island through the city of Kobe, which in itself has a population of about 1.5 million. Strong ground-shaking lasted for 20 seconds and caused severe damage over a large area.

Nearly 5500 [later raised to 6300] deaths have been confirmed, with the number of people injured reaching about 35 000. Nearly 180 000 buildings were badly damaged or destroyed, and officials estimate that more than 30 000 people were homeless on the night of the earthquake.

This life loss caused by the earthquake was the worst in Japan since the 1923 Great Kanto earthquake, when about 140 000 people were killed, mostly by the post-earthquake conflagration. The economic loss from the 1995 earthquake may be the largest ever caused by a natural disaster in modern times. The direct damage caused by the shaking is estimated at over Y13 trillion (about US$147 billion). This does not include indirect economic effects from loss of life, business interruption and loss of production.

Other indicators of the extent of the damage include:

- One-fifth of Kobe's population (310 000) were made homeless. One million households had no water.
- Several highways collapsed including 500 metres of the elevated Hanshin expressway which was tipped at a 45-degree angle [**2.10**].
- Over 150 fires broke out, mainly due to gas leaks. Calm conditions prevented conflagrations.
- Thousands of bodies lay in schools and sports centres as there was a shortage of coffins.
- The port of Kobe was devastated to the extent that 90% of the 187 berths were destroyed. Shipping was disrupted for many months.
- Many people suffered considerable financial losses as only 3% of the residents had earthquake insurance.
- Reconstruction costs could be as much as $100 billion.
- The emotional stress suffered by the survivors may create long-term social and behavioural problems.

Well water warned of Kobe quake

Peter Hadfield, Tokyo

Groundwater beneath the city of Kobe may have given out several warning signals prior to the devastating earthquake of 17 January, according to Japanese researchers.

Changes in the composition of groundwater have long been considered as possible warnings of large earthquakes. George Igarashi of Hiroshima University had begun monitoring the groundwater near Kobe only three months before the earthquake struck.

Igarashi was not expecting an earthquake. He simply wanted to gather information on the concentrations of radon in groundwater and chose Kobe because its groundwater – which is extracted from wells for use in the brewing of sake – had been extensively studied in the past.

Radon is formed from the radioactive decay of uranium in the Earth's crust. Minute quantities seep into deep groundwater from uranium-bearing rocks. If stress in the rocks opens up fissures and cracks, then more radon can leak into the groundwater. Because rocks can become stressed immediately before an earthquake, a sudden increase in the concentration of radon-222 in groundwater may herald a quake.

In late 1993 Igarashi and his colleagues made a few initial measurements of the concentration of radon in water from a well 30km north-east of the epicentre of the Kobe quake. On 27 October 1994, when the radon levels were already slightly higher than those recorded in 1993, the team began monitoring the well continuously. On 7 January 1995, radon levels surged to around 12 times their 1993 levels, before falling back to slightly below the 1993 figures on 10 January. The researchers were unable to find any explanation for the results other than seismic activity.

Radon in Kobe's groundwater

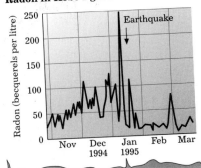

Figure 2.13 An early warning

Source: *New Scientist*, 15 July 1995

Figure 2.14 Japan's tectonic setting

Japan lies on the eastern fringe of the Eurasian continental plate. From Tokyo northwards, the Pacific oceanic plate is descending (subducting) below Japan at a rate of about 10cm a year. This has created a trench of up to 8km in depth that runs parallel to the eastern coastline. South of Tokyo, the Pacific plate is being subducted beneath the Philippine plate, which is itself subducted beneath southern Japan. The town of Kobe is situated in a geologically complex area near the northern tip of the Philippine plate.

Earthquakes occur in descending plates as they are gradually compressed by their continuous relative motion. Friction prevents smooth, continuous movement along fault planes, and forces gradually accumulate, to be released periodically and often dramatically, as the shock waves of earthquakes.

The source of an earthquake, known as its focus, is shallower the closer it is to a trench. The most destructive earthquakes tend to be associated with shallow, high-energy foci.

Because of its unusual tectonic setting, the seismic activity in the Kobe region is a combination of thrust faulting, where one block is forced on top of another along a shallow-dipping plane, and strike-slip faulting.

Source: From David Rothery, 'Kobe: anatomy of an earthquake', *Geographical Magazine*, April 1995

Architecture and engineering

- Most deaths occurred in houses that could not withstand the shaking. Ninety per cent of the fatalities were caused by the 'pancaking' effect as upper storeys collapsed onto lower storeys. Better foundations and the use of supports and bracing would help buildings respond better in an earthquake.
- Large, modern steel-framed office blocks performed well and did not collapse. Some innovative designs did perform badly, suggesting the need for more research and testing.
- Developments (housing and industry) on alluvium and other soft soils, particularly on the waterfront, were severely damaged as a result of liquefaction.
- Older bridges and elevated highways made of reinforced concrete were severely damaged but so were some recent 'earthquake-proof' structures. More research in design is necessary.
- Gas supplies were disrupted for several weeks and gas leaks caused fires. Research is needed into automatic shut-offs.
- Lack of water reduced the capabilities of the fire fighters. Leaks were widespread and led to a reduction in the water pressure in the pipes.

Preparedness and response

- Preliminary observations suggest that preparedness and emergency response efforts were less than satisfactory. The immediate urban search and rescue effort was inadequate for the thousands of buildings destroyed in this event.
- Preparedness and response are often the most affordable, if not the only possible, mitigation techniques available in many regions.
- Efforts are needed (1) to continue and increase support for emergency preparedness and response, at all levels, public and private; and (2) to encourage development of innovative techniques for improved response such as automated, rapid post-event damage assessment and decision-making using geographic information system-based tools.

Summary

There are relatively few new lessons to be learned from the Kobe earthquake from an engineering viewpoint. The real lesson is the need to motivate societies to act – to replace or strengthen deficient structures and systems, and improve planning and preparedness.

Figure 2.15 Lessons to be learned from the Kobe earthquake

Enquiry

1. ■ When did the earthquake occur?
   ■ What was its magnitude?
   ■ What was it like to experience the earthquake?
   ■ How many people died or were injured?
   ■ What were the short-term and long-term effects?

2. ■ What caused the earthquake?
   ■ Which plates were involved?
   ■ What happened at the plate margin?

3. ■ How did the characteristics of the earthquake (e.g. magnitude, depth of focus, etc.) contribute to the effects?
   ■ How important was building design?
   ■ Did people know what to do?
   ■ Were the people of Kobe and the authorities prepared for the response?

4. What lessons were learned about:
   ■ building design
   ■ preparedness (of people and authorities)
   ■ service provision?

5. How has the earthquake helped scientists in their attempts to forecast earthquakes?

6. What suggestions would you make for a city of a similar size so that any future earthquake would have a less disastrous effect?

Volcanoes

Introduction

Volcanoes are among the most dramatic and awesome reminders that the Earth is a living and evolving planet. For many thousands of years people have settled on the slopes of volcanoes benefiting from the fertile soils that develop on their flanks as lavas are rapidly weathered. Yet every few hundred years, these mountains have been split apart or, in some cases, completely blown up by volcanic eruptions. Farming communities and even whole towns have been destroyed by ash and lava. The world's climates have often been affected as millions of tonnes of ash are blasted into the atmosphere.

Today many of the world's major cities – including Naples in Italy, Tokyo and Kagoshima in Japan, Mexico City (Mexico) and Quito, the capital of Ecuador – lie perilously close to active volcanoes. Many millions of people in South America, Japan and Indonesia are at risk from the direct effects of volcanic eruptions and all of us are, in one way or another, at risk from the indirect effects.

In this chapter we examine the range of volcanic hazards, their causes and the measures being adopted to predict eruptions and reduce their impact on people's lives.

There are several different hazards associated with volcanic eruptions.

Lava flows

Comparatively speaking, lava flows do not cause anything like as much death and destruction as one might imagine. Few lava flows extend much beyond 10km from the volcanic crater. Their speed of flow is determined by the silica content of the lava. Silica-poor lava is very fluid and flows faster and further than silica-rich lava which tends to be much thicker. A lava flow may destroy farmland, buildings and lines of communication, but lives are rarely lost.

Ash

Fine-grained ash together with larger **pyroclastics** ('fire rocks') may be ejected into the air during a violent eruption. While much of it will be deposited close to the volcano, winds may spread the deposit to affect areas many kilometres away. The depth of deposits of ash can vary from as much as several metres close to the volcano to a few centimetres further away. There are several hazards associated with ash.

- The sheer weight of deposited ash can cause the roofs of buildings to collapse.
- Air thick with hot ash can lead to the asphyxiation of humans and animals. The AD79 eruption of Vesuvius and the subsequent burial of Pompeii under 3 metres of ash caused 2000 deaths, mostly from ash asphyxiation. A contemporary account said:

  The darkness was so complete that it might be compared to a sealed room in which the lamp had been put out. It was much more unnerving than that though, for they were not in a sealed room but out in the open with the air full of cries from the crowd – screams of terror and prayers for deliverance. The ash fall became heavier, piling up around them, so that every now and again they had to shake themselves free from being buried.

- Vast areas of land become blanketed with ash, wiping out crops, blocking roads and rendering all maps completely useless.
- Ash may combine with water to form mudflows called lahars.
- Ash sent high into the air may influence the global weather machine as high-level winds transport it around the Earth (**3.1**). Ash particles will block out incoming radiation (sunlight) thereby causing the Earth's surface to become slightly cooler. When Krakatau erupted in 1883, there was a 10–20 per cent decrease in solar radiation received at the Earth's surface.
- Ash will increase the supply of condensation nuclei in the lower atmosphere so promoting the formation of water droplets and increasing rainfall amounts. The abnormally wet June of 1982 was blamed on the eruption of El Chichon volcano in Mexico (**3.1**).
- During the last 15 years about 80 commercial jets have been damaged when flying through ash and several have nearly crashed.

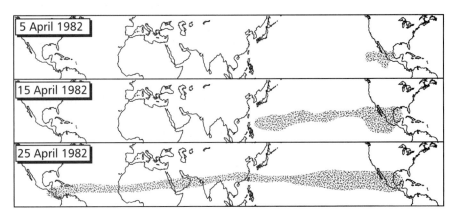

Figure 3.1 The spread of ash from El Chichon in April 1982

Pyroclastic flows and nuées ardentes

Perhaps the most devastating of all volcanic hazards is the cloud of incandescent gas, ash and rocks known as a **pyroclastic flow**. This can reach temperatures of up to 800°C and is capable of speeds in excess of 200km/hour. A pyroclastic flow like the one that which occurred at Mount Unzen (Japan) in 1991, is usually triggered by the collapse of part of the volcano's summit, often comprising a lava dome that has built up during several eruptions. Broken rock combines with newly released gases to sweep down the volcano's flanks obliterating all in its path. During the eruption of Mount St Helens (USA) in 1980, the north face of the volcano collapsed to unleash a blast that felled fully grown trees up to 25 km away.

An event that involves mostly gases, often in the form of a fireball, is called a **nuée ardente**. It was a **nuée ardente** that killed 28 000 people in St Pierre on the island of Martinique when Mt Pelée erupted in 1902. The city itself was reduced to charred wood and twisted metal in just three minutes.

Gas emissions

If you have ever visited a volcanic area you will have noticed the strong smell of sulphur in the air. In addition to sulphur, several other deadly gases can be emitted including carbon dioxide and cyanide.

Lahars

A lahar is a thick, cement-like mudflow consisting of volcanic ash and water – often melted snow and ice that once capped the volcano. It is capable of flowing at tremendous speeds of up to 100km/hour, and reaching a distance of up to 300km. Although a lahar is not a 'primary product' of a volcanic eruption, it is certainly far more devastating than, for example, a lava flow. Lahars most commonly follow river valleys, burying houses and their occupants, blocking roads and destroying bridges, and blanketing fertile floodplains with thick mud which hardens like concrete.

Jokulhlaups (glacial outburst floods)

If a volcano erupts beneath an ice cap, a vast amount of ice can be melted, and when this water eventually escapes (it often remains ponded up for some time), tremendous quantities of water and sediment will cascade down mountainsides onto the plains below. Jokulhlaups can be extremely spectacular but they rarely cause much loss of life as they usually occur in largely uninhabited areas.

In 1996 an eruption beneath the Vatnajökull glacier in Iceland melted a hole in the ice and led to water 50m deep collecting in the Grimsvotn volcanic crater. The water that eventually escaped beneath the glacier carried icebergs as high as 5-storey buildings and swept away roads and bridges to the south of the glacier.

Tsunami

Tsunami are abnormally high sea waves generated by ground shocks such as earthquakes and volcanic eruptions.

The 1883 Krakatau eruption created waves up to 35 metres high. These immense waves swept along the coasts of Java and Sumatra killing over 36 000 people. Tsunami are very often responsible for causing more loss of life and damage to property than the more direct effects of volcanic eruptions. This is because they affect low-lying, often heavily populated coastal plains some distance away from the volcano itself, which are not therefore thought to be under immediate threat from eruptions. Too often, local people are often completely unprepared for the waves.

If any proof were needed of the lethal nature of volcanic eruptions, **3.2** lists the most deadly of those known in history.

Deaths	Volcano	Date	Major cause of death
92 000	Tambora, Indonesia	1815	Starvation
36 417	Krakatau, Indonesia	1883	Tsunami
29 025	Mt Pelée, Martinique	1902	Ash flows
25 000	Ruiz, Colombia	1985	Mudflows
14 300	Unzen, Japan	1792	Volcano collapse, tsunami
9350	Laki, Iceland	1783	Starvation
5110	Kelut, Indonesia	1919	Mudflows
4011	Galunggung, Indonesia	1882	Mudflows
3500	Vesuvius, Italy	1631	Mudflows, lava flows
3360	Vesuvius, Italy	79	Ash flows and falls
2957	Papandayan, Indonesia	1772	Ash flows
2942	Lamington, Papua New Guinea	1951	Ash flows
2000	El Chichon, Mexico	1982	Ash flows
1680	Soufrière, St Vincent	1902	Ash flows
1475	Oshima, Japan	1741	Tsunami
1377	Asama, Japan	1783	Ash flows, mudflows
1335	Taal, Philippines	1911	Ash flows
1200	Mayon, Philippines	1814	Mudflows
1184	Agung, Indonesia	1963	Ash flows
1000	Cotopaxi, Ecuador	1877	Mudflows
800	Pinatubo, Philippines	1991	Roof collapses and disease
700	Komagatake, Japan	1640	Tsunami
700	Ruiz, Colombia	1845	Mudflows
500	Hibok-Hibok, Philippines	1951	Ash flows

Figure 3.2 The most deadly volcanic eruptions

1 How true is it to say that lava is the main product and the main danger of a volcanic eruption?

2 What are the direct hazards associated with ash?

3 Study **3.1**, which shows the spread of ash following the eruption of El Chichon (Mexico) in 1982. Describe the spread of the ash cloud and estimate its rate of movement.

4 What is a pyroclastic flow and how does it differ from a nuée ardente?

5 Suggest why lahars can be so devastating to communities both in the short term and the longer term. Why is their effect so much greater in developing countries?

6 A tsunami is an indirect effect of a volcanic eruption, yet it can cause massive loss of life. Explain why this is so.

7 Study **3.2**.
a Which country has experienced the greatest number of the most deadly eruptions?
b What appears to be the most common cause of death?
c Suggest reasons for starvation being the main cause of death in two of the eruptions listed.
d Which country appears to most severely affected by tsunami? Why is this?
e Were death tolls heavier in the past compared with recent times? Draw a graph plotting deaths against the date, to discover if recent eruptions have caused fewer deaths. (You could use vertical bars to represent the deaths drawn on a timeline, or you could plot the points as a scattergraph. Leave out the AD79 eruption of Vesuvius – this will make your graph easier to construct.)

SECTION B

Case study: Nevado del Ruiz, 1985

Nevado del Ruiz is one of a chain of active volcanoes in the northern part of the Andes in Colombia (**3.3**). It rises to 5200 metres. Up until 1985 it had been capped by ice and had several glaciers on its flanks. The volcano began to threaten a major eruption long before it actually occurred. During 1984 there were several small earthquakes and steam eruptions and, by the end of February 1985, the crater lake had turned into sulphuric acid! During September and October there were several small eruptions and much melting of the ice, causing river levels to be abnormally high.

Shortly after 9pm on 13 November the volcano exploded into life, sending a column of ash and pyroclastics some 8000 metres into the air. Burning ash and rocks fell onto the ice and the combination of meltwater and ash formed deadly lahars which swept down the river valleys engulfing all in their path. Shortly after 10pm one lahar reached the town of Armero, where in a matter of seconds it buried the 22 000 inhabitants in up to 8 metres of mud.

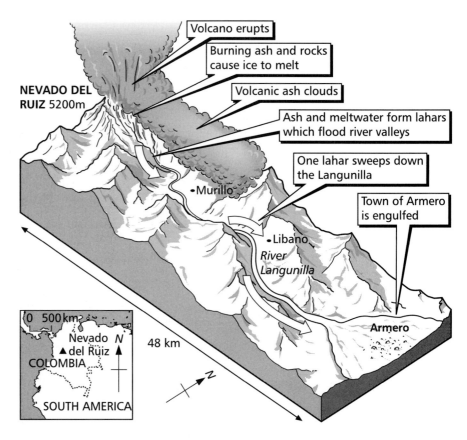

Figure 3.3 The eruption of Nevado del Ruiz

Volcano erupts

Burning ash and rocks cause ice to melt

Volcanic ash clouds

Ash and meltwater form lahars which flood river valleys

One lahar sweeps down the Langunilla

Town of Armero is engulfed

NEVADO DEL RUIZ 5200m

•Murillo

•Libano

River Langunilla

Armero

0 500km

Nevado *N* ▲ del Ruiz
COLOMBIA

SOUTH AMERICA

48 km

The correspondent for *The Guardian* described the scene immediately after the lahar hit Armero:

> *It looks like a beach at low tide, just mud and driftwood. Trees, houses and cars were all carried off. Few houses were still standing and the roofs are crowded with people waiting to be rescued.*
>
> *It was very difficult to reach the scene for all the rivers were swollen and many bridges were down. There are hundreds of corpses floating down the river and a suffocating smell of sulphur in the air.*
>
> *It has to be one of the world's biggest mass graves. A few relatives can be seen picking through the muck and others gaze down in disbelief. One 12-year-old girl, trapped in the mud from the waist down, said she was standing on the corpses of her father and aunt.*

Case study: Mount Pinatubo, 1991

In June 1991 Mount Pinatubo on the island of Luzon in the Philippines exploded into life, having remained dormant for some 600 years. It was one of the most dramatic eruptions in recent years, and had significant effects both on the local people and on the world.

■ 800 local Filipinos lost their lives. They were farmers and workers living on the slopes of what, for them, had been a 'safe' volcano, worshipped as the home of their god.

- Two-thirds of the dead were killed as a result of catastrophic lahars which raced down the inhabited valleys burying whole villages and their inhabitants.
- 650 000 people lost their livelihoods as 80 000 hectares of farmland were covered by metres of ash and mud.
- 1.2 million people lost their homes and 500 000 people migrated to the already overcrowded capital city of Manila.
- 4.25 cubic miles of rock, ash and gas were ejected 15 000 metres into the atmosphere (by comparison Mount St Helens ejected only 0.25 cubic miles), causing a cooling of the planet by approximately 0.5°C as radiation from the Sun was blocked out and reflected back into space. Figure **3.4** shows the impact on surface air temperatures.

Figure 3.4 The effect of the Mount Pinatubo eruption on global surface air temperatures

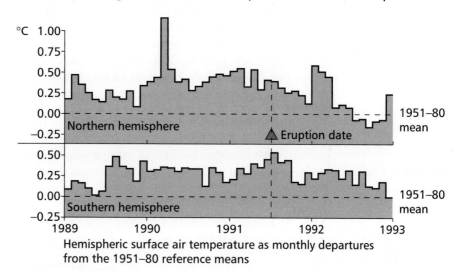

Hemispheric surface air temperature as monthly departures from the 1951–80 reference means

Mount Pinatubo did 'warm up' prior to the main eruption and hundreds of thousands of people were evacuated. However, the long-term effects of the eruption – loss of homes, farmland, jobs and livelihoods – are very serious and will be felt for many years to come in what is one of the poorest parts of the world.

Review

8 Referring to the text and **3.3**, write a short report on the effects of the eruption of Nevado del Ruiz. What, if anything, could have been done to reduce the loss of life?

9 Identify the short-term and longer-term effects of the eruption of Mount Pinatubo in 1991.

10 Study **3.4**.
  a For the northern hemisphere, describe the general trend of temperature before and after the eruption, using actual figures to support your description.
  b Did the eruption of Mount Pinatubo affect temperatures in the southern hemisphere as much as in the northern hemisphere? Use an atlas to explain your answer.
  c What does **3.4** tell us about the effect of volcanic eruptions on global climate? Try to give reasons for the effects you have identified.

What causes volcanic eruptions?

A volcanic eruption occurs when molten magma rises and escapes at the ground surface. Almost all volcanoes occur at plate margins (*Environment and People*, figure **2.4** page 12), with a particular concentration around the edge of the Pacific Ocean – the so-called 'ring of fire'.

At a destructive plate margin, particularly where continental crust is involved as in North and South America, magma tends to be rich in silica (andesitic) and very viscous in its consistency. As the magma is so viscous, it often solidifies before reaching the surface, forming a plug in the volcano. This acts like a cork in a champagne bottle and can lead to the build-up of tremendous pressure within the volcano itself. Eventually this pressure is released in the form of a violent explosion. Such eruptions emit vast amounts of ash and rock, and *nuées ardentes* are common. Lahars may be formed as ice melts and the resultant water combines with ash and landslide debris to form lethal mudflows. Recent examples of this type of eruption include Mount St Helens (USA) 1980, Nevado del Ruiz (Colombia) 1985, Mount Unzen (Japan) 1991 and Mount Pinatubo (Philippines) 1991.

Figure 3.5 Volcanoes and plate tectonics – a summary

The magma that feeds volcanoes at constructive plate margins and **hot spots** (isolated pockets of magma away from plate margins) contains far less silica and is, as a result, far more fluid. As it rises to the surface, the greater fluidity enables gas bubbles to expand, so preventing the sudden

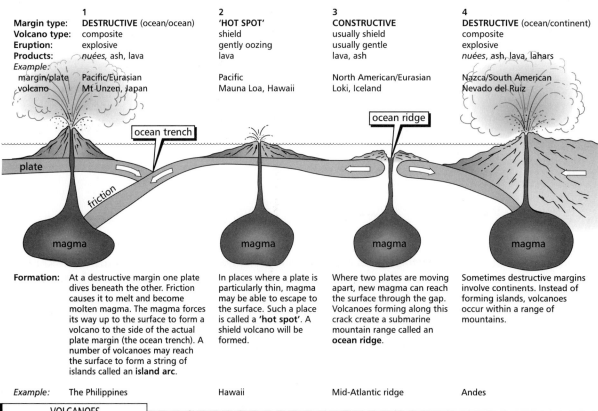

	1	**2**	**3**	**4**
Margin type:	DESTRUCTIVE (ocean/ocean)	'HOT SPOT'	CONSTRUCTIVE	DESTRUCTIVE (ocean/continent)
Volcano type:	composite	shield	usually shield	composite
Eruption:	explosive	gently oozing	usually gentle	explosive
Products:	nuées, ash, lava	lava	lava, ash	nuées, ash, lava, lahars
Example:				
margin/plate	Pacific/Eurasian	Pacific	North American/Eurasian	Nazca/South American
volcano	Mt Unzen, Japan	Mauna Loa, Hawaii	Loki, Iceland	Nevado del Ruiz

ocean trench

plate

friction

ocean ridge

magma

magma

magma

magma

| **Formation:** | At a destructive margin one plate dives beneath the other. Friction causes it to melt and become molten magma. The magma forces its way up to the surface to form a volcano to the side of the actual plate margin (the ocean trench). A number of volcanoes may reach the surface to form a string of islands called an **island arc**. | In places where a plate is particularly thin, magma may be able to escape to the surface. Such a place is called a **'hot spot'**. A shield volcano will be formed. | Where two plates are moving apart, new magma can reach the surface through the gap. Volcanoes forming along this crack create a submarine mountain range called an **ocean ridge**. | Sometimes destructive margins involve continents. Instead of forming islands, volcanoes occur within a range of mountains. |
| *Example:* | The Philippines | Hawaii | Mid-Atlantic ridge | Andes |

surface explosive activity associated with thicker magmas. Eruptions tend to be fiery but generally far less explosive than those at destructive margins, and their consequences tend to be less catastrophic.

Figure **3.5** summarises the relationship between volcanoes and plate tectonics. Whilst there is a fairly clear relationship between a volcano's type and its eruptive characteristics, and the type of plate margin on which it lies, it is important to stress that every volcano is unique.

- No two magma chambers are identical in terms of chemical composition, water and gas content, depth, size, etc.
- The speed at which magma rises to the surface will vary. This influences the change in pressure and the subsequent expansion of gases within the magma.
- The resistance and shear stress of the overlying rocks will vary from one volcano to another. Whilst the vent is usually the weakest part of the volcano resulting in vertical eruptions, this is not always the case. The eruption of Mount St Helens in 1980 involved a lateral blast due to the collapse of the northern flank of the volcano.

Although it is true to say that the majority of the most damaging eruptions have involved destructive margin volcanoes, it would be misleading to suggest that destructive margin volcanoes are necessarily any more hazardous to people than constructive margin volcanoes.

Constructive volcanoes erupt more frequently so the hazard is more 'real' to people. Frequent flows of lava and falls of ash render large areas sterile – such areas cannot realistically be farmed or settled and so they have very low population densities. Destructive margin volcanoes, on the other hand, erupt far less frequently – there can be many hundreds of years between eruptions – so the hazard becomes less 'real' to people. Memories are short and people think that the risk is minimal, so they settle and farm areas that are, tectonically and geologically speaking, potentially dangerous. The elements weather the volcanic rock to produce fertile soils which further attract settlers to these areas, and as a result population densities are often high.

Review

11 Explain why volcanic eruptions at destructive plate margins are more explosive and violent.

12 Contrast the hazards associated with volcanoes at destructive plate margins with those at constructive plate margins.

13 What are hot spots?

SECTION D

Predicting volcanic eruptions

Unlike earthquakes, the prediction of volcanic eruptions has been more successful as volcanoes often give warning signs of an imminent eruption. For example, the eruption of Mount Unzen (Japan) on 3 June 1991 followed many months of activity, enabling evacuations to take place so that only 43 people lost their lives compared with some 15 000 when the same volcano erupted in 1792. Several techniques are used by scientists to monitor volcanoes for signs of activity.

Geophysical monitoring

This involves studying the shape of the volcano and detecting bulges or increased tilt which might be indicative of rising magma. In 1980, for example, a very noticeable bulge appeared on the north side of Mount St Helens just a few weeks prior to the eruption. As the bulge grew steadily, so it became increasingly obvious to scientists that the volcano was going to erupt, and an evacuation was ordered.

Scientists in Hawaii have perfected instrumentation capable of measuring the deflation or inflation of volcanoes to an accuracy of 0.0000573 degrees. This is the equivalent of placing a 10p coin under a plank of wood 2km long. Hawaii has several volcanoes which erupt very frequently, allowing scientists to perfect their technique. At Kilauea, tilt increases as magma fills an underground reservoir (**3.6**). The tilt continues to increase until the outer rocks can no longer contain the pressure of the rising magma and an eruption occurs. After the eruption, the summit deflates.

Satellites are becoming increasingly valuable in monitoring changes in volcano shape. Satellites of the Global Positioning System (GPS) send signals to receivers positioned on the flanks of volcanoes. The signals enable positions to be measured to the nearest few centimetres, so giving a detailed and up-to-the-minute picture of any changes in a volcano's topography.

Figure 3.6 Predicting eruptions by changes in the angle of tilt

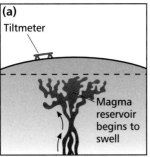
(a)
Tiltmeter
Magma reservoir begins to swell

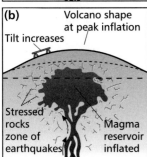
(b)
Volcano shape at peak inflation
Tilt increases
Stressed rocks zone of earthquakes
Magma reservoir inflated

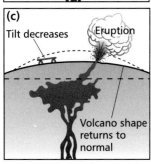
(c)
Tilt decreases
Eruption
Volcano shape returns to normal

The angle of tilt changes as magma fills the chamber.

Increased seismic activity

An increase in earthquake frequency and the migration of foci towards the volcano itself can be a very good indication of an imminent eruption. These earthquakes tend to be increasingly shallow in depth and are usually the result of rupturing of rock as magma rises.

Gravitational changes

Rising magma often causes a localised increase in gravitational pull as dense material rises to the surface. The Hawaiian Volcano Observatory has a network of gravity stations in operation constantly monitoring the active volcanoes on Big Island, one of the Hawaiian chain.

Geochemical changes

It is generally believed that rising magma will lead to changes in any one of a number of gas and water characteristics, e.g. concentrations of sulphur and carbon dioxide, or changes in water temperature.

Scientists from the Los Alamos National Laboratory in New Mexico, USA have carried out extensive studies of Kilauea and Mauna Loa on Hawaii. They believe that small changes occur in the concentrations of trace elements (such as gold, platinum and iridium) in gases and particles emitted by the volcanoes in the days prior to an eruption.

Whilst scientists continue to monitor gases and liquids in the hope of discovering that small changes link up with eruptions, it is worth noting that no such changes were identified before Mount St Helens erupted in 1980.

Thermal energy monitoring by satellite

Recently, satellites run by the USA's National Oceanic and Atmospheric Administration (NOAA) have carried a device called an Advanced Very High Resolution Radiometer (AVHRR) which is capable of measuring infrared radiation given off by volcanoes. Although this research is at a very early stage, it may well prove to be a valuable addition to the ground monitoring systems currently employed.

Whilst all this painstaking monitoring may well suggest that an eruption is likely to occur 'soon', there is as yet no foolproof technique capable of giving an accurate time of an eruption. In addition, it has proved very difficult to accurately predict the magnitude of an eruption and, therefore, suggest which areas should be evacuated and where people should be moved to. It is a sobering fact that 12 out of the 16 largest and most deadly eruptions of the last 200 years (**3.2**) have been from volcanoes that were considered to be inactive.

Review

14 To what extent do you think that people's perception of the volcanic hazard varies depending upon their quality of life and the economic status of the country in which they live?

15 Referring to **3.6**, explain how changes in the angle of tilt can be used to predict eruptions.

16 What do scientists look for when studying the geochemistry of volcanoes?

SECTION E

Reducing the volcanic hazard

There are a number of different actions that can be taken to reduce the effect of volcanic eruptions.

Controlling flows

Lava flows can be managed quite successfully and controlled by erecting barriers to divert free-flowing lava away from valuable land or property; such barriers have been widely used on Hawaii. A lava flow can be bombed to break it up. This encourages it to cool more rapidly as a greater surface area is exposed to the air. In 1983, explosives were used to divert a lava flow on Mount Etna that was threatening nearby villages.

Cooling the lava fronts with water encourages lithification (rock formation). This strategy was very successful in arresting the advance of lava erupting on the island of Heimaey (Iceland) in 1973 when it seemed as if the harbour of the main port Vestmannaeyjar was going to be blocked off. Hoses directed 6 million tonnes of sea water for five days onto the lava. Although this strategy was considered a national joke – *pissa a hraunid* (a *hraunid* means 'on the lava', and you can probably guess the meaning of *pissa*!) – the harbour entrance, though reduced in width, remains open to this day, thereby guaranteeing the livelihoods of the local fishermen.

It is also possible to use barriers to direct and, if needs be, divert lahars away from populated areas. In the Philippines, earth and rock banks have been constructed to contain lahars in certain valleys so as to prevent them spilling over into populated or heavily cultivated valleys.

Building design

The danger of building collapse can be avoided to some extent by encouraging appropriate building design involving sloping roofs to discourage the formation of thick and weighty deposits of ash.

Exposure (hazard) mapping

If there is sufficient historical evidence such as documented records of past eruptions, or geological evidence such as deposits of lava and ash, it is possible for scientists to construct a map showing the likely danger areas in the vicinity of a volcano. The map produced is called an **exposure** or **hazard map** and it indicates those areas at greatest potential risk from an eruption. Recent advances in radar have enabled digital elevation models (DEMs) to be produced. These volcanic models can be used to suggest the likely paths of lava and pyroclastic flows.

Figure 3.7 Exposure map of Soufrière volcano (Guadeloupe)

Figure **3.7** is an exposure map for the volcano Soufrière on the Caribbean island of Guadeloupe. Notice that it shows how various parts of the island are at risk from different volcanic hazards such as volcanic bombs (large

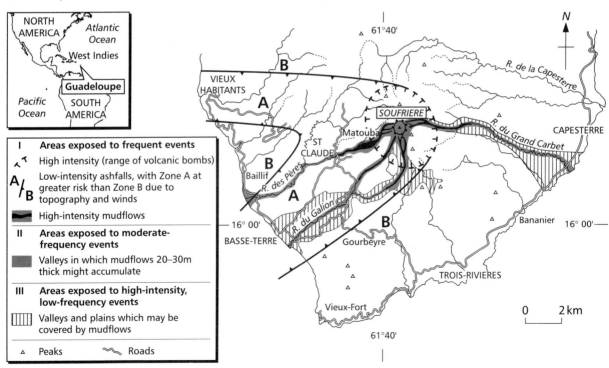

I	Areas exposed to frequent events
High intensity (range of volcanic bombs)
Low-intensity ashfalls, with Zone A at greater risk than Zone B due to topography and winds
High-intensity mudflows
II	Areas exposed to moderate-frequency events
Valleys in which mudflows 20–30m thick might accumulate
III	Areas exposed to high-intensity, low-frequency events
Valleys and plains which may be covered by mudflows
△ Peaks   Roads

lumps of rock), ash and lahars. Notice also the way in which the map takes account of different intensities of eruption and the effects of topography and wind direction.

It is interesting to note that in 1976 the entire population of Guadeloupe was evacuated when the eruption of Soufrière seemed imminent. The volcano did not erupt at all and there was considerable controversy afterwards about whether the evacuation was appropriate. The exposure map (**3.7**) was a response to the debate –it was published shortly afterwards.

As volcanic eruptions vary so much, no exposure map could ever hope to be 100 per cent accurate. All it can do is give planners a general guide, in much the same way as weather forecasts predict our weather. If the map is based on inadequate or inaccurate evidence, it might do more harm than good, particularly if what were thought to be 'safe' areas prove to be otherwise.

Review

17 Describe three measures that can be taken to control the advance of a lava flow.

18 a What is an exposure map and what purpose does it serve?
  b What sort of evidence can be used by scientists when trying to construct such a map?
  c Should people be prevented from living close to volcanoes?

19 Study **3.7**.
  a The area immediately surrounding Soufrière volcano is vulnerable to 'volcanic bombs'. What are they?
  b In which direction is ash expected to be a particular hazard? What does this tell you about the prevailing winds in Guadeloupe?
  c The expected paths of lahars are very concentrated yet tend to extend a long way from the volcano. Why is this?
  d Comment on the likely value of such a map to:
  ■ the government and planners in Guadeloupe
  ■ the ordinary peasant farmers living on the volcano's flanks or farming the fertile valley floors.

W E B SEARCH

The United States Geological Survey (USGS) and the Cascades Volcano Observatory have excellent sites and there is information about preparing for eruptions of Cascade Mountain volcanoes. Search also within Volcano World and the Asian Urban Disaster Preparedness Center (www.adpc.ait.ac.th).

Extended enquiry: Montserrat

The tiny Caribbean island of Montserrat hit the headlines in July 1995 as its dormant volcano, Soufrière Hills, burst into life. Two years later numerous eruptions threatened to wipe out the British dependency, and many of the island's human features – its houses, hospitals and communications – were destroyed.

Figure 3.8 Diary of a volcano – Soufrière Hills

July 1995
Having lain dormant for some 350 years, the Soufrière Hills volcano bursts into life with an eruption of ash and pyroclastics. Much of the south of the island is covered in a thick layer of ash. The capital Plymouth has to be evacuated – it becomes a ghost town.

June 1997
The volcano erupts again, sending pyroclastic flows down its flanks and ash clouds 10 000 metres into the air. The fiery eruptions destroy seven villages and kill 19 people.

3 August 1997
Pyroclastic flows reach Plymouth, igniting fires which threaten to destroy the city. There are ashfalls in Isles Bay, Old Towne and Salem. The UK government considers evacuating the entire population of Montserrat.

6 August 1997
Further violent eruptions. Pyroclastic flows reach Plymouth, virtually destroying the capital. Eighty per cent of buildings are badly damaged or totally destroyed. Ash deposits fall across the island including in the 'Safe Zone' in the north. 1.4 metres of ash now covers Plymouth. The previously considered safe zone in the centre of Montserrat has now been evacuated, including the towns of Salem, Hope, and Olveston.

25 August 1997
Scientists warn of a cataclysmic eruption. Only 4500 of the 11 000 inhabitants are left on the island, most having left for neighbouring islands such as Antigua. Many of those who remain feel that the UK Government has not offered them enough compensation for their losses. Adults have been offered US$4000 and children $1700. By remaining on the island they hope to increase pressure on the UK Government.

30 August 1997
The north side of the volcano collapses, triggering pyroclastic flows and a large ash cloud.

9 September 1997
The Montserrat Volcanic Observatory urges all residents to move to the far north of the island as a further eruption appears to be imminent.

The UK Government announces that it will rebuild on the north of the island, and $3840 is offered to anyone who wants to relocate off the island.

21 September 1997

A large pyroclastic flow buries part of the airport runway and sets fire to the terminal building. The UK Government promises $64 million to rebuild houses, hospitals and the airport.

23 September 1997

The UK Government considers using force to move people out of the evacuation zone. By holding a live concert in London, Elton John, Eric Clapton and others raise US$1.6 million for relocation and new housing.

25 September 1997

Rock fragments and ashfalls affect Old Towne and St John's.

2 October 1997

Three explosions send ash clouds over 12 000 metres into the air. Pyroclastic flows reach the sea through Tar Valley and White River valley. Ash and rock fragments fall right across the island.

26 December 1997

Dome collapse leads to another huge eruption. Pyroclastic flows and ash and rockfalls affect the White River valley and the village of St Patrick's.

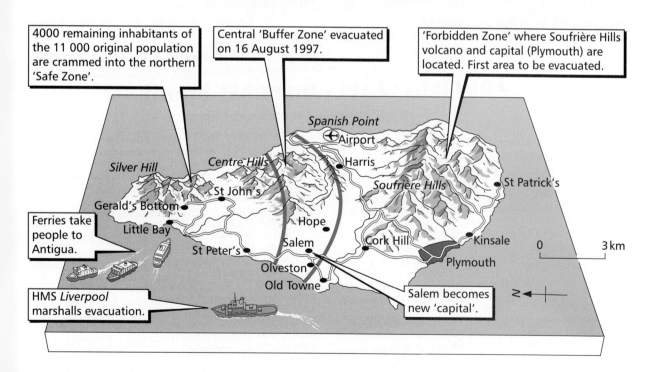

4000 remaining inhabitants of the 11 000 original population are crammed into the northern 'Safe Zone'.

Central 'Buffer Zone' evacuated on 16 August 1997.

'Forbidden Zone' where Soufrière Hills volcano and capital (Plymouth) are located. First area to be evacuated.

Ferries take people to Antigua.

HMS *Liverpool* marshalls evacuation.

Salem becomes new 'capital'.

Spanish Point
Airport
Silver Hill
Centre Hills
Harris
St John's
Soufrière Hills
St Patrick's
Gerald's Bottom
Little Bay
Hope
St Peter's
Salem
Cork Hill
Kinsale
Olveston
Plymouth
Old Towne

0   3 km

Figure 3.9 Evacuation of Montserrat

Phil Davison
reports from Olveston
as the islanders face up to 'voluntary evacuation'

It could have been a Fifties English film, with Alec Guinness as the colonial governor, but the crowd in front of him was deadly serious. Above them, black smoke and ash billowed from the Soufrière Hills volcano. A few hundred yards offshore, the frigate HMS *Liverpool* lay at anchor, ready to help evacuate the remaining 4000 inhabitants of Montserrat.

The volcano has gradually squeezed these people into a small area in the north of this British Caribbean island. The protesters were angered by what they consider the British government's vacillation and lack of clarity over whether to evacuate the island and, if so, how much assistance they should receive.

Angry and frustrated over their emergency living conditions and confused about a 'voluntary evacuation' offer, they banged drums, and marched to the British Governor's office to demand his resignation.

'In the event of a total evacuation, we want to make it clear that we are not abandoning our country but expect to return here when it is safe to do so,' said group spokeswoman Teresa Silcott as the Governor listened. If Britain did not respond, she said, Montserrat, one of a dozen British Dependent Territories, would demand independence.

The Governor laid out the package on offer. First, those wishing to evacuate to Britain would be put up in hotels in nearby Antigua, fed three meals a day and transported to Britain within about a week at Britain's expense. Second, he supported a package put forward by Chief Minister Osborne the night before, under which a family of four would receive £27 500 over a period of 18 months as evacuation compensation. Britain was considering the proposal, he said. Third, Britain would support anyone who remains on the island.

It is not only the people of Montserrat who are angry. The neighbouring Caribbean islands, too, consider that Britain has been inactive. Antigua, which has already seen its own population of 65 000 swollen by the arrival of more than 4000 Montserratians over the past two years, is now asking for financial aid to deal with the refugees. 'We want to be helpful to the Montserratians and to the British government but at the same time the British government must help us,' said Ronald Sanders, High Commissioner of Antigua and Barbuda in London. 'There are now as many Montserratians on Antigua as on Montserrat. We've almost reached breaking point.'

Figure 3.10 Anger over evacuation

Source: *The Independent*, 21 August 1997

Figure 3.11 Looking for disasters

Tourism chiefs forecast boom
By **Harvey Elliott** travel correspondent

OFFICIALS on Montserrat plan to attract large numbers of 'volcano tourists' to the devastated island once the immediate threat of further eruptions is over.

Montserrat was one of the least known Caribbean islands until the volcano made international news. Now local tourism chiefs believe that the island can benefit from the publicity by encouraging holidaymakers – particularly Americans – to watch the explosions and clouds of steam and ash from a safe distance. One possibility is to run cruise ships close by, enabling passengers to watch the lava-flow through binoculars, or to fly sightseeing tours above the volcano.

Volcano watching, although not attracting a huge following, does already have hundreds of devotees. Derek Moore, operations director of the British tour operator Explore Worldwide,

said there was a small but regular demand for organised visits to seven main volcano areas around the world.

'They are only small groups and the most popular is a visit to three volcanoes in Sicily,' he said. 'We send 12 groups each year to see Stromboli and others on the island and generally get around 16 people in each.'

Other destinations include Ecuador, Nicaragua, Iceland, the Azores and Crete.

'People are interested in the historical effects volcanoes can have on a country. Even though they may be extinct many people find themselves fascinated to see at close hand how the Earth's geology is made up,' Mr Moore said.

'The problem with Monteserrat, however, is that it is extremely dangerous and we cannot have people risking their lives.'

In this 'Extended enquiry' you should draw upon information taken from **3.8–3.11** and any other sources you can find to investigate the following aspects.

- The nature of the volcanic hazards caused by the eruptions.
- The effect of the eruptions on the people of Montserrat.
- The response of the UK Government.
- The success of predicting eruptions based on experiences in Montserrat.
- The long-term implications of the eruptions on the people and on the country.

You should aim to cover the points listed above in the form of a report. To help you in its preparation, you should first consider the following questions and activities.

Enquiry

1
- What were the types of hazard associated with the eruptions of the volcano?
- How many people had been killed by the end of 1997?
- Draw a sketch map of the island to plot the locations of the hazards. (This is not as easy as it sounds – be prepared to draw a rough version first. Mark the three zones – 'forbidden', 'buffer', and 'safe' – on your map.)

2
- What effects did the eruptions have on the people of Montserrat? Consider the uncertainties involved, the need to be evacuated, the encouragement to leave for another island, etc.

3
- Construct a timetable to show the response of the UK Government to the ever-increasing scale of the disaster.

- How did the local people react to the help offered?
- How appropriate and successful are 'superstar' pop concerts such as the one held in London in September 1997?

4
- What warning signs of impending eruption did the scientists study?
- Given the number of casualties, how successful were the scientists and the local authorities in preventing human disaster on a massive scale?

5
- Assess the future for Montserrat. What could/should be done for the future?
- How might tourism provide some much-needed income?
- Has the UK Government an obligation to rebuild on the island?

WEB SEARCH

The volcano became less active in the early part of 1998. However, there is every likelihood that some activity will have taken place since this book was written. Search for the 'Montserrat Volcanic Observatory'. Look also for 'Volcano World' which has an enormous amount of material on volcanoes.

Landslides

A landslide is a sudden downhill movement of part of a slope. It is often unexpected and can cause considerable damage to property and lines of communication, as well as injury and loss of life. It is, however, important to remember that landslides are perfectly natural processes and most of the time they are quite harmless, occurring in remote and often mountainous regions. There are many different types of landslide, but geomorphologists usually distinguish between **heave**, **slide** and **flow** (*Environment and People*, figure **3.3** page 24; figure **3.4** in the same book shows in a triangular graph how these relate to each other).

SECTION A

What causes landslides?

There are usually several factors that combine to cause a slope to collapse or fail. It is worth distinguishing between those factors that can make a slope unstable (**4.1**) and those that actually trigger a landslide.

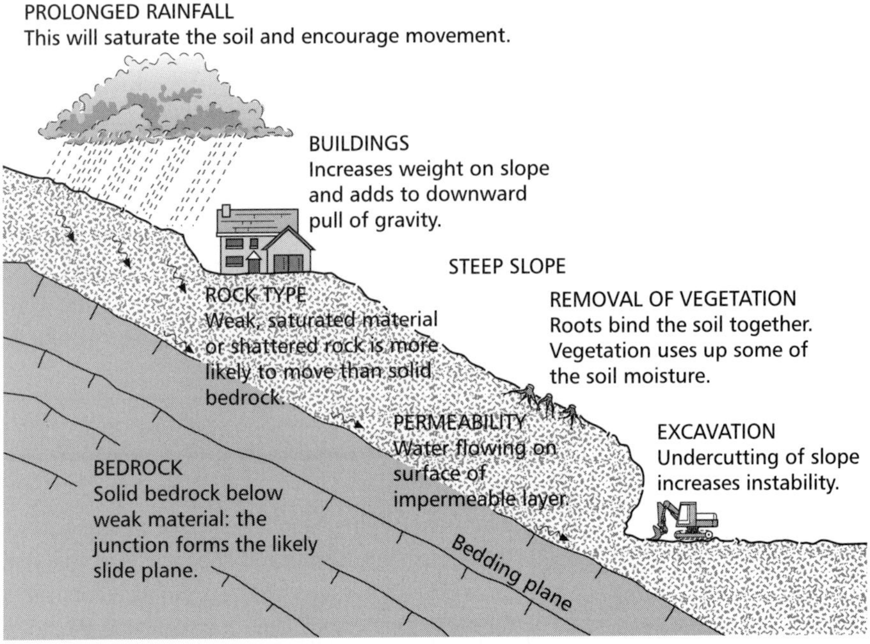

PROLONGED RAINFALL
This will saturate the soil and encourage movement.

BUILDINGS
Increases weight on slope and adds to downward pull of gravity.

STEEP SLOPE

ROCK TYPE
Weak, saturated material or shattered rock is more likely to move than solid bedrock.

REMOVAL OF VEGETATION
Roots bind the soil together. Vegetation uses up some of the soil moisture.

PERMEABILITY
Water flowing on surface of impermeable layer

EXCAVATION
Undercutting of slope increases instability.

BEDROCK
Solid bedrock below weak material: the junction forms the likely slide plane.

Bedding plane

Figure 4.1 Factors contributing to slope instability

The following factors contribute to slope instability (**4.1**).

- **Slope gradient** The steeper the slope, the more likely it is to collapse. If a slope is made steeper by undercutting, either naturally by the action of a river or the sea or unnaturally by human excavation, it is more likely to fail.

- **Rock type** Rocks have different strengths and abilities to retain a slope angle. Tough rocks such as granite or schist can be stable at very steep angles whereas weaker sands and clays can slip downhill at the gentlest of angles. Some rocks are permeable, letting water pass through them, whereas others are impermeable and unable to transmit water. If water can pass through one rock but not the one beneath it, it might pond-up underground and lubricate a landslide.
- **Geological structure** If lines of weakness such as bedding planes (the junctions between beds of sedimentary rocks), faults and joints are aligned downslope they encourage slope collapse.
- **Waterlogging** Water held within rock and soil can exert considerable outwards pressure thus weakening the cohesion of material on a slope. The saturation of a slope is one of the most frequent causes of slope failure.
- **Drought** Very dry conditions, particularly in generally wet environments, can lead to shrinkage of clays and the formation of cracks. Both can weaken the material and make it more prone to collapse, particularly after a period of heavy rain.
- **Slope loading** Buildings and other developments can exert a downward force on a slope, making it more likely to move.
- **Deforestation** Cutting down trees and removing vegetation can make a slope much more vulnerable to collapse. The roots of plants help hold soil together and plants also use up some of the water that might be present in a slope. Furthermore, lack of vegetation cover means that rainfall will no longer be intercepted. Interception slows down the impact of a storm on a slope.
- **Weathering** Tropical regions, with high temperatures and heavy rainfall, often have deeply weathered soils which are very weak and unstable on a slope and are prone to landslides.

Most landslides result from a combination of the factors described above – see the case studies in Section B.

Trigger mechanisms

If a slope is potentially unstable, it may only need a very minor additional circumstance to trigger movement and collapse. There are three particularly common trigger mechanisms.

- **Heavy rainfall** If a slope is already saturated and lubricated, heavy rain will result in a sudden surge of water which may overcome friction and cause the slope to collapse. This is probably the most widespread cause of slope failure.
- **Earthquakes** It is easy to understand how an earthquake, in shaking the ground, will trigger landslides. Very often it is landslides that cause the greatest loss of life during an earthquake. Apart from simply

Review

1 What is a landslide?

2 Identify the mechanisms that trigger landslides. For each one, explain how they can lead to slope failure.

dislodging material, earthquake tremors can cause some materials, for example silts, to become jelly-like and extremely vulnerable to collapse. This is **liquefaction** (see page 15).

■ **Basal excavation** In addition to making a slope potentially unstable, excavation at the base of a slope may also be responsible for triggering a landslide. In 1988, 66 people perished at Catak in Turkey when a newly-created 55° slope collapsed following the excavation of a road cutting.

SECTION B

Case study: Vaiont Reservoir (Italy)

On 9 October 1963, 2043 people died when a huge slab of rock slid 400 metres into the Vaiont Reservoir in the Italian Alps to the north of Venice causing a huge wave some 100 metres high to top the dam and surge through several villages in the valley below (**4.2**).

Figure 4.2 The Vaiont disaster, 1963

(a) Map of the Vaiont area

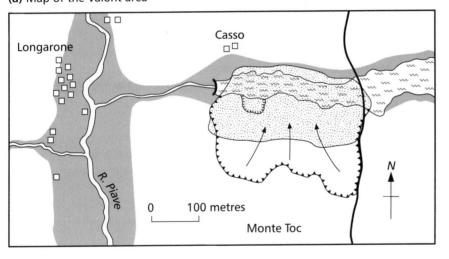
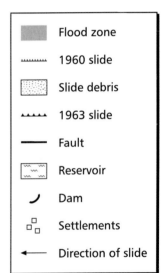

▓	Flood zone
⋯⋯	1960 slide
⬚	Slide debris
⋏⋏⋏	1963 slide
—	Fault
〜	Reservoir
⌣	Dam
⬚	Settlements
←	Direction of slide

(b) Geological factors contributing to the Vaiont landslide

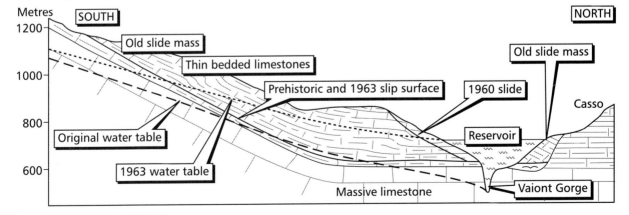

Review

3 Figure **4.2** includes a cross-section across the Vaiont valley. Make a copy of the diagram and add labels to show the following factors that contributed to the disaster:

- steep slope
- saturated rock caused by rising water table as the reservoir filled
- slip surface
- ancient landslide comprising weakened rock resting on solid limestone bedrock.

4 With the aid of sketches, identify the factors that probably contributed to the Thredbo landslide of 1997.

Figure **4.2** shows that several factors contributed to the landslide. The upper part of the slope that collapsed was itself an ancient landslide, now resting on solid limestone. This site should never have been chosen for a reservoir, given the unstable nature of the slope. The reservoir, which began filling in 1960, caused an increase in the amount of underground water and increased water pressure in the rocks. The trigger for the landslide was probably heavy rainfall.

Case study: Thredbo (Australia)

On 30 July 1997, a landslide destroyed two ski-lodges in the Australian resort of Thredbo (**4.3**). The landslide occurred without warning in the middle of the night. A steep embankment of the Alpine Way collapsed, sending tonnes of mud and rock downhill. The landslide crashed into the two-storey Carinya Lodge and this in turn was sent crashing into the Bimbadeen Lodge below. Eighteen people lost their lives but, miraculously, one person was pulled out alive after three days of being trapped in freezing conditions.

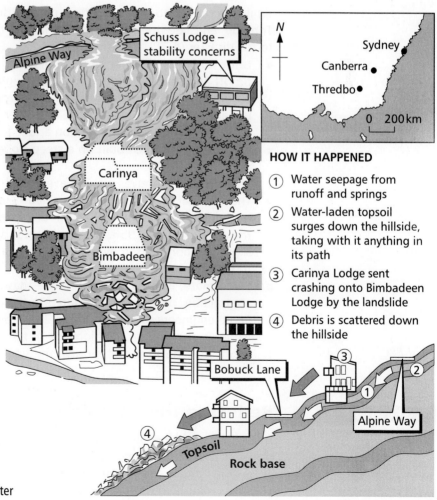

HOW IT HAPPENED

① Water seepage from runoff and springs

② Water-laden topsoil surges down the hillside, taking with it anything in its path

③ Carinya Lodge sent crashing onto Bimbadeen Lodge by the landslide

④ Debris is scattered down the hillside

Figure 4.3 The Thredbo disaster

Eyewitness accounts describe the event.

Ⓐ *After the whoosh of the wind and the crashing of collapsing buildings, came the screams. They lasted just a few minutes then there was the silence, almost as brief. Next the commotion of people emerging from the night to witness an awesome and terrible sight. Yet there was not so much a sense of crisis or danger as one of helplessness, a frantic search for purposeful activity . . . a call for more torches to search the rubble.*

Ⓑ *I had just gone to bed and heard something almost like an explosion. You could hear all the crashing and banging . . . all the buildings sliding down, and all the cars too. The horns went off and people were screaming. It could have been seconds, it might have been minutes – I have no idea. It was horrific. So I just jumped out of bed, grabbed some clothes and ran outside . . .*

It was pitch black. You could hear people shouting and the people in my lodge then came outside. We were trying to get torches and you could smell gas and petrol everywhere. There were people under the rubble and we knew exactly where they were. We watched them for three hours . . . I mean, they were under the rubble but we knew where they were.

Whilst there was no obvious trigger mechanism involved in the Thredbo landslide, there were several factors that probably contributed to the disaster (**4.3**).

- Large amounts of water had accumulated underground as a result of runoff and springs. Water had in fact begun to bubble-up into the basement of the Carinya Lodge in the days before the disaster. That underground water would have lubricated the upper surface of the rock, thereby overcoming friction and leading to the landslide.
- The lodges had been built on a steep slope with a gradient of more than 1 in 4 (25%).
- The lodges themselves would have added extra weight to the slope.
- The material on which the lodges had been constructed may have contained loose debris (called **fill**) discarded during the building of the Alpine Way in 1956. The road was subject to frequent small slips, probably caused by poor drainage, and was constantly having to be repaired.
- The climate, with its cycles of freeze–thaw, expansion and contraction, may have contributed to the disaster.
- Some engineers suggested that a 2-metre-high concrete wall built earlier in the year on the high side of the road adjacent to the slip might have diverted water down the slope.

Reducing the landslide hazard

In an ideal world, people should live as far away from potentially unstable slopes as possible. However, this is not always possible. As populations and cities grow, land previously unoccupied becomes settled. The need to construct roads and railways through mountainous areas, to extract resources or grow food, further impinges on marginal land that would be best left alone. As technology increases, so it is possible to manage slopes to reduce the potential hazards.

There are various strategies that can be adopted to increase slope stability and make collapse or failure less likely.

Reducing the slope gradient

Material can be built up at the bottom of a slope – this is called **toe loading**. It has the effect of reducing the gradient of the slope, thereby making it more stable. A variety of materials can be used such as loose rock fill, concrete or earth banks.

Improving drainage

Water building up within a slope is one of the main causes of slope failure. Drainage systems designed to prevent such a build-up of water are a common and effective hazard-reducing measure. Figure **4.4** identifies some methods used by engineers. Notice that these measures are intended to reduce water entering the slope as well as to encourage any water that does get into the slope to drain away quickly.

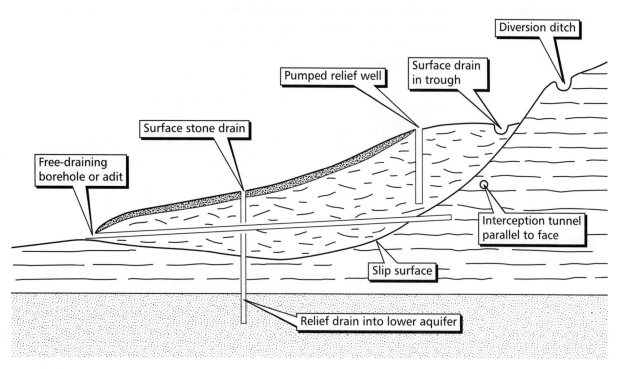

Figure 4.4 Methods of improving slope drainage

Supporting the existing slope

There are a huge number of engineering solutions aimed at securing slopes (**4.5**). These include the use of nets to prevent falling rocks, steel cables and sprayed concrete (called **shotcrete**) to anchor rock surfaces, and rock bolts to retain isolated boulders.

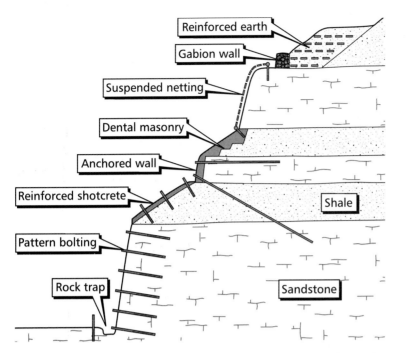

Figure 4.5 Methods of slope stabilisation

Labels on figure:
- Reinforced earth
- Gabion wall
- Suspended netting
- Dental masonry
- Anchored wall
- Reinforced shotcrete
- Pattern bolting
- Rock trap
- Shale
- Sandstone

Land use management

Overgrazing and deforestation can be causal factors in slope collapse. Vegetation reduces the impact of rain, and plants use up some of the water as they grow. In addition, plant roots help bind together loose soil and rock, increasing slope stability. Careful slope management aimed at reducing animal grazing can help to make a slope more stable, and afforestation has the same effect.

Hazard mapping

Potentially hazardous slopes can be identified using satellite and field surveys, and landslide hazard maps can be drawn. Such maps are in frequent use in the Alps to locate areas at risk from avalanches. One of the most important pieces of information is the location of past landslides, often indicated by hummocky ground or scars on a hillside.

Landslide monitoring

Slopes that are known to be unstable can be monitored (**4.6**). Whilst this may seem very reassuring, it is interesting to note that the Vaiont slide was monitored . . . but it failed unexpectedly!

Another possible line of 'action' is, of course, to do nothing! This might be regarded as appropriate if the costs of stabilisation outweigh the benefits. It might, for example, be cheaper for a local authority to do nothing and just clear away any loose rocks as and when they fall. Such a solution is, though, only really acceptable when there is no great danger to life or property.

Figure 4.6 Monitoring slope stability

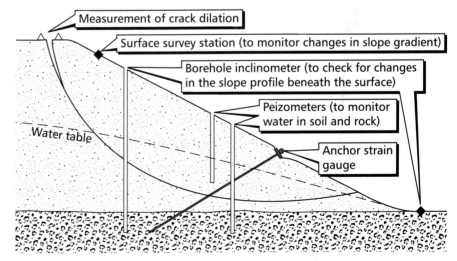

Labels on figure:
- Measurement of crack dilation
- Surface survey station (to monitor changes in slope gradient)
- Borehole inclinometer (to check for changes in the slope profile beneath the surface)
- Peizometers (to monitor water in soil and rock)
- Anchor strain gauge
- Water table

5 Which of the drainage methods shown in **4.4** do you think would be most cost-effective? Justify your choice.

6 Referring to **4.5**, compare shale and sandstone slopes in terms of:

■ their potential instability
■ the best ways of improving their stability.

7 Study **4.6**. For each monitoring method, explain why it is appropriate for indicating any slope instability.

SECTION D

Extended enquiry: Pokfulam landslide, Hong Kong

In this activity you will investigate the causes of a serious but not unusual landslide. Landslides similar to this are very common throughout the world and it is important that you study a typical landslide rather than one of the more dramatic and less common disasters.

Write a report on the Pokfulam landslide addressing the following points.

■ The sequence of events leading up to the landslide.
■ The factors contributing to the landslide.
■ The landslide trigger.
■ Measures for making similar slopes more stable.
■ Lessons for the future.

To help you in the preparation of your report, you should first study the information below and then attempt to answer the questions that follow.

Report on the landslide

On 8 May 1992 a landslide occurred on a hillslope above Baguio Villas in Pokfulam, Hong Kong. Landslides are common in Hong Kong due to the very steep slopes and the heavy rainfall. This particular landslide involved the sudden collapse, at about 2pm, of a masonry retaining wall, which released loose and saturated debris (it had been raining heavily for two days) onto a steep slope. Further collapse of the debris (above where the wall used to be) occurred shortly after 3.25pm. The debris then flowed down a steep gully before crashing into Lower Baguio Villas, killing a child and a government engineer (**4.7**).

Several factors contributed to the collapse of the slope. There had been very heavy rainfall for two days. On the morning of 8 May, 180mm of rain fell in just two hours and a further 40mm fell in 20 minutes at around 2pm. Nearly 300mm of rain had fallen since 6am (by comparison, the average **annual** rainfall of London is about 600mm). This intense rainfall, too heavy for the drains to cope with, soon saturated and weakened the soil. Surface runoff resulting from the heavy rain may well have been responsible for finally triggering the collapse of the retaining wall.

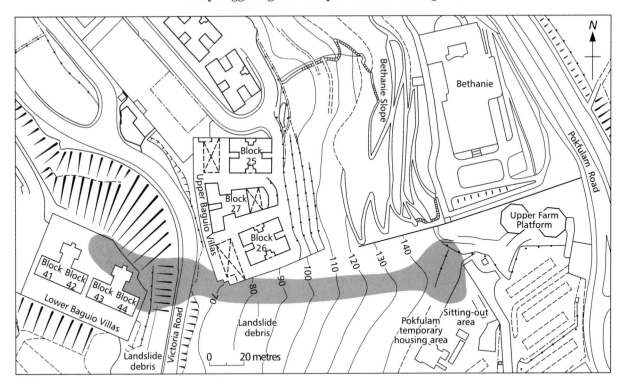

Figure 4.7 Topography of the hillside and the extent of the landslide

The wall itself was constructed many years ago and did not have the benefit of modern engineering methods. Poor drainage and the possible collapse of material in front of the wall may have contributed to its failure. Once breached, the loose material behind the wall simply flowed down the hill (**4.8**). This fill material was possibly discarded during the development of the site as a farm in the 1880s. It was later flattened to form a sitting-out platform for local residents.

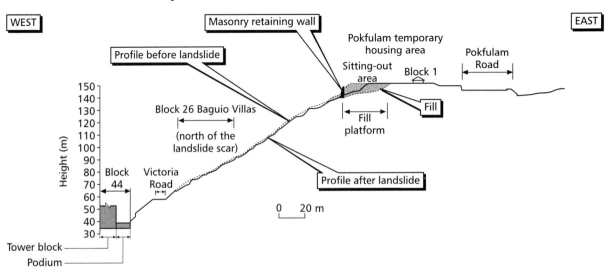

Figure 4.8 Cross-section of the hillside at the landslide location

Below the wall, a normally dry gully helped to channel the saturated rock and soil downslope towards the Baguio Villas. As the material moved it eroded the gully further, thereby increasing its volume.

Enquiry

1 ■ Describe the sequence of events that led to the landslide.
   ■ Use **4.7** to establish the dimensions of the landslide (length, slope, width, etc.).
   ■ Use **4.8** to draw a simple-cross section to show what happened.

2 Assess the importance of the following in contributing towards the landslide:
   ■ rainfall
   ■ the retaining wall
   ■ the nature of the material behind the retaining wall
   ■ the characteristics of the slope (steepness, presence of a gully, etc.).

3 Explain what triggered the collapse of the wall that led to the landslide.

4 ■ Using your knowledge of mitigation methods, suggest what might be done to stabilise other slopes with similar characteristics to this one at Pokfulam.
   ■ What lessons were learned at Pokfulam that might help prevent similar events occurring?

WEB SEARCH

The National Landslide Center in the USA can be accessed via the United States Geological Survey (USGS). Try:

//geohazards.cr.usgs.gov/html_files/nlicsun.html

to discover excellent fact sheets.

Material on the Thredbo disaster may still be available – try a search on: Thredbo+landslide.

Another site worth trying for information on landslides and mudflows is at: //marauder.millersv.edu/~ehem/LANDSLIDES.html

For information on Hong Kong, locate the 'Bibliography on the Geology and GeoTechnical Engineering of Hong Kong' at: www.info.gov.hk/ced

For information on avalanches (to be covered in a future text in the EPICS series) see the Scottish Avalanche Information Service for some excellent brochures. See also 'Cyberspace Snow and Avalanche Center' at: www.csac.org/

5

Severe weather

In this chapter, we move from geological and geomorphological hazards to those associated with weather and climate. Most weather and climate hazards are created by both extremes of temperature (heat and cold) and precipitation (wet and drought). With reference to air movement, only one extreme is relevant: very strong wind. Strong winds are a characteristic linking all three hazards investigated in this chapter: tropical cyclones, tornadoes, and extreme winter weather.

Tropical cyclones

What is a tropical cyclone?

Tropical cyclones are very intense depressions or areas of low pressure that bring heavy rain and extremely strong winds. They are also known as **typhoons** in eastern Asia and the western Pacific Ocean, as **hurricanes** in the Caribbean and Atlantic Ocean, and as **cyclones** in the Indian Ocean.

They form over seas in the tropics in the summer and then move across the Earth's surface at a speed of about 15–30 km/h. They tend to follow well-recognised courses (**5.1**).

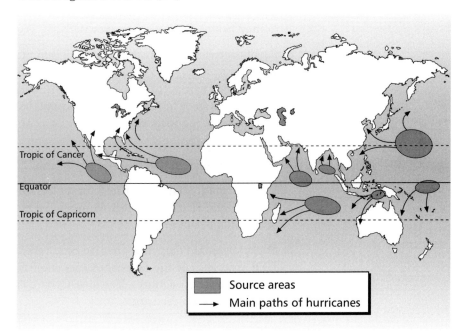

Figure 5.1 Tropical cyclone source areas and common tracks

1 Study 5.1. On an outline world map, plot and label the following:
- the main source areas of tropical cyclones
- the common tracks taken by tropical cyclones
- the countries/ regions at greatest risk from cyclone damage.

2 Of the factors leading to the development of a tropical cyclone, which do you think are the most important? Justify your choice.

3 Explain why tropical cyclones die when they reach land.

Cyclones can be tracked using satellites and radar. This makes it possible to forecast their movement and to issue warnings to coastal areas judged to lie in the forthcoming path of a cyclone. However, despite the monitoring, tropical cyclones are generally considered to cause more loss of life than any other natural hazard. Some 15 per cent of the world's population live in coastal areas under threat from tropical cyclones and, on average, 15 000 people are killed by them every year.

The precise reasons for the formation of a tropical cyclone are still not fully understood, although there are certain conditions that seem to encourage their formation.

- They always form over oceans, from which they derive their energy. The condensation of water vapour releases latent heat, so increasing the intensity and power of the cyclone.
- They seem to begin as a cluster of thunderstorms centred around a localised area of low pressure within the **intertropical convergence zone**.
- Sea surface temperatures need to exceed 26.5°C to create the required amount of latent heat to 'fuel' the cyclone.
- The atmosphere needs to be sufficiently unstable to enable the formation of deep cumulonimbus clouds.
- A high level of moisture is required, hence their formation over the sea.
- They always form in a location of more than 5° north or south of the Equator, as rotation of the system will not occur if the Coriolis force is zero (i.e. at the Equator).

A tropical cyclone typically measures 300–500km across and, when photographed from a satellite, it looks like a catherine-wheel spiral of cloud (*Environment and People*, figure **11.20** page 143). It moves over the sea picking up heat and moisture but then dies out on reaching cooler waters or land. It is the heat and the moisture of the warm ocean that keeps it going.

What are the effects of tropical cyclones?

Several hazards are associated with tropical cyclones. These are most frequently experienced in low-lying coastal areas, such as Bangladesh, or on islands, such as those in the Caribbean. The hazards include the following.

- **Very strong winds**, often in excess of 200km/h. Such winds will destroy homes, flatten crops and trees, damage overhead power lines, and whisk up paving slabs and roofing tiles to form lethal flying weapons. In the Caribbean, strong winds can devastate tree crops such as bananas, on which several countries depend for a large proportion of their income. For example, in 1979 Hurricane Allen destroyed over 90 per cent of the banana plantations on St Lucia and St Vincent.
- **Very heavy rainfall** which can exceed 800mm in a single day. Such intense rainfall often leads to flooding which can cause considerable damage to property and loss of life in the days and weeks following a tropical cyclone event.

Figure 5.2 A storm surge

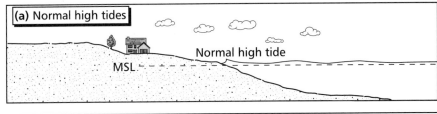

(a) Normal high tides

Normal high tide

MSL

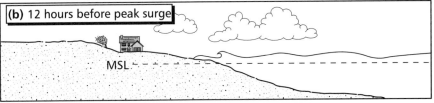

(b) 12 hours before peak surge

MSL

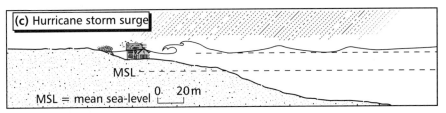

(c) Hurricane storm surge

MSL

MSL = mean sea-level   0   20m

■ **A storm surge (5.2).** This is a raised dome of water typically 60–80km across and up to 8m above normal high tide. It is by far the most devastating effect of a tropical cyclone as it can wipe out whole coastal communities. It is caused by a combination of the very low pressure having a 'suction effect' on the sea, and the extremely strong winds. In 1970 a storm surge of 7m swept over 26 000km² of low-lying Bangladesh killing over 225 000 people and leaving 2.5 million homeless. In 1992, Hurricane Andrew caused tremendous damage along the south-east coast of Florida. On that occasion 65 people were killed, 250 000 were made homeless and damage was estimated at well over US$25 billion (*Environment and People*, page 142).

Review

4 a Outline the typical short-term effects of a tropical cyclone.
  b Suggest some longer-term effects.
  c Compare the effects of cyclones in developed countries with those in developing countries.

5 a What is a storm surge?
  b Explain why storm surges are often responsible for the greatest destruction during a tropical cyclone.
  c Make a copy of **5.2** and add labels to describe what is happening.

Reducing the tropical cyclone hazard

The most common method of mitigation or adjustment involves the accurate tracking of tropical cyclones using satellites and radar.

Warnings can then be issued to communities that seem to be under threat. There are four levels of warning and advice (**5.3**). If necessary, and feasible, evacuation can be ordered.

Figure 5.3 Advice and measures relating to tropical cyclones

ADVISORIES
Tropical depression advisory
Provides information on the development and threat of a tropical depression which becomes a threat to land areas. The system is not named unless it is a hurricane or tropical storm that has been down-graded to a tropical depression. Each new tropical depression is assigned a number, however.

Tropical storm advisory
Issued when the wind speed of a tropical cyclone reaches 63km/h or higher. Tropical storms are given names.

Hurricane watch
Advisory issued for a particular area when conditions are favourable for the development of a hurricane. It does not necessarily mean that a hurricane is imminent.

Hurricane warning
Issued when hurricane conditions are expected to affect a particular area within 24 hours.

PRECAUTIONARY MEASURES
On hearing advisory
- Continue normal activities but stay tuned to radio and television for further messages.

On hearing hurricane or tropical storm watch
- Review emergency preparedness requirements, especially family emergency plans.
- Continue to listen to weather advisories on radio.
- Be ready to take quick action in case of a warning.
- Establish contact points.

On hearing hurricane or tropical storm warning
- Stay tuned to the radio for information.
- Protect property and personal possessions (including important documents).
- Place indoors, loose objects found in and around the yard.
- Fill up car with gasoline [petrol].
- Pick fruit and trim trees if near house.
- Store water, food and essential medicines.
- Feed animals and pets and move them indoors or let them loose.
- Know where you are going to shelter if the need arises.

Source: Caribbean Disaster Emergency Response Agency (CDERA)

Review

6 Tropical cyclones are a natural hazard that to some extent can be forecast.
 a How is that forecasting undertaken?
 b What are the four levels of advice that can be issued to the public?

7 Study 5.3. Suggest reasons for each of the precautionary measures to be taken on the issuing of a hurricane warning.

8 What measures other than warnings might be adopted to mitigate tropical cyclone damage, particularly in coastal regions?

Property can be protected from storm damage, and sea walls constructed to guard against the effects of storm surges. In some countries, shelters have been built on higher ground to offer a refuge from storm surges. However, in low-lying and densely-populated countries, such as Bangladesh, there is only a limited amount that can be done by means of a response to any warnings issued.

The Caribbean Disaster Emergency Response Agency (CDERA) has produced a list of actions that could be taken to reduce the vulnerability of buildings and infrastructures.

- New buildings should be designed to be wind- and water-resistant.
- Communications and power lines should be installed underground in coastal areas.
- Buildings should be raised off the ground to reduce damage from storm surges.
- Mangroves should be planted along the coast to hold together sand and reduce the impact of a storm surge.
- River embankments should be built and maintained to prevent overtopping following heavy rain.

Case study: Tropical cyclone in Bangladesh

Bangladesh is frequently hit by tropical cyclones and, since much of the country is low-lying, damage is often significant.

On 19 May 1997, the flat coastal region of Chittagong was hit by a severe tropical cyclone with winds up to 200km/h and a storm surge of 3–4m. According to government estimates, 100 people were killed and more than 7000 injured. Nearly 2000 head of cattle were lost. The cyclone affected nearly 2 million people and caused serious damage to more than half a million houses, more than 600 educational buildings, 29 000ha of crops and 152km of embankments. Compared with earlier cyclones, there was very little loss of life. This was because the cyclone hit at low tide during the day, and also because many people had already been safely evacuated to cyclone shelters situated on higher ground.

Review

9 Explain why:
- Bangladesh suffers from tropical cyclones
- the death toll of the 1997 event was not as high as in earlier cyclones.

SECTION B

Tornadoes

Nature and causes

A tornado is a violent windstorm often associated with thunderstorms or tropical cyclones. It is characterised by a narrow, twisting, funnel-like column of cloud that reaches down to the ground from a huge cumulonimbus storm cloud (5.4). Tornadoes rarely touch ground for more than 20 minutes but they can do so several times from the same cloud over a period of time.

A tornado is usually no wider than 100m across, but the wind speeds associated with it are enormous, often exceeding 480km/h. Tornadoes carve a destructive path that may be less than 500m across and rarely over 20km in length. The winds are capable of lifting cars and mobile homes, uplifting trees and removing roofs from houses. Loose objects, such as wood and bricks, are hurled through the air and are often responsible for any injuries or loss of life. The US National Weather Service reports that on average 42 people are killed by tornadoes in the USA every year. The Fujita tornado scale is commonly used to measure tornadoes (5.5).

Figure 5.4 Tornado in Jasper, Minnesota

Tornadoes can occur at any time of the year. In the USA they are most common in the summer in the Midwest and the South. The soaring daytime temperatures set off thunderstorms due to intense convective activity. Over 80 per cent of US tornadoes occur in the afternoon. In the UK, where they are surprisingly common though generally weak, tornadoes tend to be a winter phenomenon associated with small but intense areas of low pressure.

Figure 5.5 The Fujita tornado scale

F-0:	*65–115km/h*	Chimney damage, tree branches broken
F-1:	*116–180km/h*	Mobile homes pushed off their foundations or overturned
F-2:	*181–250km/h*	Considerable damage, mobile homes demolished, trees uprooted
F-3:	*251–325km/h*	Roofs and walls torn down, trains overturned, cars thrown
F-4:	*326–415km/h*	Well-constructed walls levelled
F-5:	*416–510km/h*	Homes lifted off their foundations and carried considerable distances, cars thrown as far as 100m

Source: Federal Emergency Management Agency (FEMA)

The precise reasons for the formation of a tornado are not fully understood, although a number of factors seem to increase the likelihood of their formation:

- the prevalence of moist and potentially very unstable air
- a significant temperature gradient at altitude, such as at a cold front (cold air undercutting warmer air)
- intense convection resulting from the land heating up during a summer's day
- the presence of cumulonimbus or thunderstorm clouds.

Reducing the hazard

Mitigation involves a combination of forecasting and community preparedness. In the USA the National Weather Service issues warnings when the weather conditions are favourable for tornado formation. In addition, stormchasers (enthusiasts who follow and track tornadoes) make reports of sightings, which can be used to warn communities. The Internet is increasingly being used for the exchange of information on tornado locations.

As tornadoes tend to be clustered in certain favoured regions, people can be prepared for them and instructed what to do. The safest place is indoors in a basement – some properties have storm cellars specifically designed as tornado refuges. Figure 5.6 lists the safety rules published by the US National Weather Service.

TORNADO SAFETY RULES

In homes, the basement offers the greatest safety. Seek shelter under sturdy furniture if possible. In homes without basements, take cover in the central part of the house, on the lowest floor, in a small room such as a closet or bathroom, or under sturdy furniture. Keep away from windows.

In shopping centres, go to a designated shelter area (*not* to your parked car).

In office buildings, go to an interior hallway on the lowest floor, or to the designated shelter area.

In schools, follow advance plans to a designated shelter area, usually an interior hallway on the lowest floor. If the building is not of reinforced construction, go to a nearby one that is, or take cover outside on low, protected ground. Stay out of auditoriums, gymnasiums, and other structures with wide, free-span roofs.

In automobiles, leave your car and seek shelter in a substantial nearby building, or lie flat in a nearby ditch or ravine.

In open country, lie flat in the nearest ditch or ravine.

Mobile homes are particularly vulnerable and should be evacuated. Trailer parks should have a community storm shelter and a warden to monitor broadcasts throughout the severe storm emergency. If there is no shelter nearby, leave the trailer and take cover on low, protected ground.

Figure 5.6 Tornado safety rules

Case study: Tangail (Bangladesh), 1996

On 13 May 1996, a tornado with winds of 200km/h tore through the district of Tangail in northern Bangladesh. In a 20-minute period of destruction it killed 700 people, injured 33 000 and destroyed 17 000 houses in 80 villages. It snapped telephone lines, uprooted trees, killed livestock and flattened crops. The tornado was one of the most devastating in recent history.

It took 24 hours before the government sent any relief, as early reports had suggested that only 22 people had been killed. The Bangladesh Army cleared rubble, buried bodies and installed tubewells to supply 'safe' water to the villages. Mercifully, disease did not break out. The government gave households corrugated iron to repair roofs, and a cash handout. Many non-governmental agencies became involved in the relief operation. Organisations such as Oxfam, Concern and the Red Crescent constructed houses, and installed latrines and tubewells. Trees were also replanted.

Case study: Selsey (UK), 1998

During the night of 8 January 1998, a tornado struck the small West Sussex village of Selsey. Fortunately no one was killed and only two people were injured, but an enormous amount of damage was done to over 1000 buildings (**5.7**).

Figure 5.7 The Selsey tornado

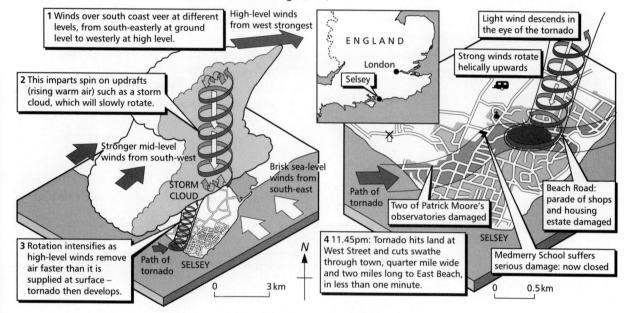

1 Winds over south coast veer at different levels, from south-easterly at ground level to westerly at high level.

2 This imparts spin on updrafts (rising warm air) such as a storm cloud, which will slowly rotate.

Stronger mid-level winds from south-west

STORM CLOUD

3 Rotation intensifies as high-level winds remove air faster than it is supplied at surface – tornado then develops.

Path of tornado

SELSEY

0   3 km

High-level winds from west strongest

ENGLAND

London

Selsey

Brisk sea-level winds from south-east

Path of tornado

N

4 11.45pm: Tornado hits land at West Street and cuts swathe through town, quarter mile wide and two miles long to East Beach, in less than one minute.

Light wind descends in the eye of the tornado

Strong winds rotate helically upwards

Two of Patrick Moore's observatories damaged

Beach Road: parade of shops and housing estate damaged

SELSEY

Medmerry School suffers serious damage: now closed

0   0.5 km

Source: *The Daily Telegraph*, 9 January 1998

The following extract is based on a report in *The Daily Telegraph*, 9 January 1998.

> *A resident of Selsey woke up at about midnight startled by the machine-gun sound of hailstones on his window. 'I was woken by this almighty sound and was convinced the window was about to cave in. The hail was so strong, the curtains were shaking. I pulled the covers over my head and hoped for the best.' The window did blow in and for a few frantic seconds the wind spun around the room. The entire gable end was then blown away. 'I could see my neighbour's house. The wall had just disappeared.'*

The Selsey tornado was a weak tornado measuring only F-1 on the Fujita scale (see **5.5**). It was caused by cold air flowing over a warm sea which triggered strong updrafts and formed an intense thunderstorm with hailstones the size of golf balls.

10 How does a tornado differ from a tropical cyclone?

11 Why are tornadoes so common in the south of the USA during summer afternoons?

12 Figure 5.8 lists the 25 deadliest tornadoes to hit the USA. Use an atlas to help you plot the locations of the tornadoes on a blank map of the USA. Devise a method to show the number of deaths. Describe the pattern that you have produced and try to account for it.

13 What factors apart from high temperatures contribute to tornado formation?

14 Study 5.6. Make a list of the main forms of advice offered and for each one explain why it is likely to reduce the hazardous impact of a tornado.

15 Why are tornadoes difficult to forecast accurately?

16 How might the Internet and the hobby of stormchasing help mitigate the tornado hazard?

17 Referring to 5.7, explain the formation of the Selsey tornado.

18 a According to the Fujita scale (5.5), what sort of damage might have been expected as a result of the Selsey tornado?
b How closely did this match what actually happened?
c How useful do you think the Fujita scale is?

19 a What were the effects of the Tangail tornado in 1996?
b How did each of the following organisations contribute to the relief to the area:
■ the Bangladesh army
■ the Bangladesh government
■ non-government agencies?

20 Contrast the effects of tornadoes in developed countries with those in developing countries.

Figure 5.8 The 25 deadliest US tornadoes

	Date	Location(s)	Deaths
1	18 March 1925	Missouri, Illinois, Indiana	689
2	6 May 1840	Natchez: Mississippi	317
3	27 May 1896	St Louis: Missouri	255
4	5 April 1936	Tupelo: Mississippi	216
5	6 April 1936	Gainesville: Georgia	203
6	9 April 1947	Woodward: Oklahoma	181
7	24 April 1908	Amite: Louisiana; Purvis: Mississippi	143
8	12 June 1899	New Richmond: Wisconsin	117
9	8 June 1953	Flint: Michigan	115
10	18 May 1902	Goliad: Texas	114
	11 May 1953	Waco: Texas	114
12	23 March 1913	Omaha: Nebraska	103

	Date	Location(s)	Deaths
13	26 May 1917	Mattoon: Illinois	101
14	23 June 1944	Shinnston: West Virginia	100
15	18 April 1880	Marshfield: Missouri	99
16	1 June 1903	Gainesville, Holland: Georgia	98
	9 May 1927	Poplar Bluff: Missouri	98
18	10 May 1905	Snyder: Oklahoma	97
19	24 April 1908	Natchez: Mississippi	91
20	9 June 1953	Worcester: Massachusetts	90
21	20 April 1920	Starkville: Mississippi; Waco: Alabama	88
22	28 June 1924	Lorain, Sandusky: Oklahoma	85
23	25 May 1955	Udall: Kansas	80
24	29 September 1927	St Louis: Missouri	79
25	27 March 1890	Louisville: Kentucky	76

SECTION C

Severe winter weather

A diversity of potential hazards

Whilst some of us look forward to snow and crisp, sunny days, winter weather can be extreme and quite hazardous.

Many parts of the world can suffer from extreme cold with daytime temperatures remaining well below freezing. If combined with a strong wind, the **wind-chill effect** can be such that the human body rapidly loses heat (**5.9**). Prolonged exposure to the cold can cause frostbite and hypothermia, both of which can be life-threatening. **Freezing temperatures** can also damage fruit crops and other vegetation. Pipes in houses may freeze and then burst on thawing; rivers may ice over, so interrupting navigation.

If **heavy snow** is combined with strong winds, or **blizzards**, this can cause considerable disruption. Whole cities can be paralysed as roads become blocked, and public utilities such as telephone lines and power cables are damaged. In rural areas, farms and villages may be cut off for days, and in mountains heavy snow can cause avalanches. The cost of snow removal, repairs and loss of business can have a significant economic effect on a city.

One of the worst hazards is **freezing rain**. This is not frozen rain as such but rather it is the result of rain falling onto surfaces that are below freezing. The rain freezes on contact with such surfaces. The ice makes roads lethal, and the weight of ice can bring down power cables and cause branches to break off trees.

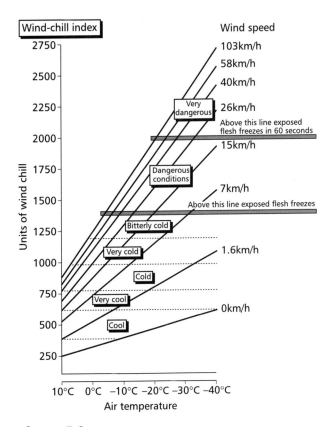

Figure 5.9
The wind-chill factor

Chart labels: Wind-chill index, Wind speed, Units of wind chill, Air temperature

Wind speed values: 103km/h, 58km/h, 40km/h, 26km/h, 15km/h, 7km/h, 1.6km/h, 0km/h

Wind-chill index scale: 2750, 2500, 2250, 2000, 1750, 1500, 1250, 1000, 750, 500, 250

Air temperature: 10°C, 0°C, −10°C, −20°C, −30°C, −40°C

Categories: Very dangerous, Above this line exposed flesh freezes in 60 seconds, Dangerous conditions, Above this line exposed flesh freezes, Bitterly cold, Very cold, Cold, Very cool, Cool

Fog is more common in winter than in summer. Low temperatures mean that water in the air will condense readily, and in calm conditions, thick fog may form. If the temperatures are low enough **freezing fog** may form. Like freezing rain, this can be a considerable hazard to motorists.

In the USA, about 70 per cent of all winter deaths occur in cars and 25 per cent are people caught out in the open. Half of those dying of exposure are over 60 years old. The National Oceanic and Atmospheric Administration in the USA offers advice to people caught in a winter storm (**5.10**). Weather bulletins advise of serious weather – you may have seen such advisories on the British TV. People likely to be affected by severe winter weather need to be prepared. Extra stocks of food, water and fuel should be obtained by householders. Cars should have emergency supplies in the form of a winter storm survival kit comprising extra clothes, sleeping bags, food, water, a compass and a shovel.

Case study: January blizzard, eastern USA

The blizzard that affected eastern USA in January 1996 was one of the worst storms of the century. Huge amounts of snow fell between 6 and 8 January. With over 500mm of snow falling on Central Park, New York was particularly badly affected. A state of emergency was declared.

The overall death toll from the blizzard was 154. A further 33 died when heavy rain combined with massive snowmelt to cause flooding. Over 200 000 properties had to be evacuated and many bridges and roads were damaged or destroyed.

The cause of the heavy snow was a deep depression bringing relatively warm and very moist air up the east coast. This air cooled when it came into contact with a large, cold air mass anchored over north-eastern USA. The depression was forced to slow down, resulting in a prolonged snowstorm.

OUTSIDE

Find shelter:

- try to stay dry
- cover all exposed parts of the body.

No shelter:

- prepare a lean-to, windbreak or snow cave for protection from the wind
- build a fire for heat and to attract attention
- place rocks around the fire to absorb and reflect heat.

Do not eat snow. It will lower your body temperature. Melt it first.

IN A CAR OR TRUCK

Stay in your car or truck. Disorientation occurs quickly in wind-driven snow and cold.

Run the motor for about 10 minutes each hour for heat:

- open the window a little for fresh air to avoid carbon monoxide poisoning
- make sure the exhaust pipe is not blocked.

Make yourself visible to rescuers:

- turn on the dome light at night when running engine
- tie a coloured cloth (preferably red) to your antenna [aerial] or door
- after snow stops falling, raise the hood [bonnet] to indicate trouble.

Exercise from time to time by vigorously moving arms, legs, fingers and toes to keep blood circulating and to keep warm.

AT HOME OR IN A BUILDING

Stay inside. When using **alternative heat** from a fireplace, wood stove, space heater, etc.:

- use fire safeguards
- ventilate properly.

No heat:

- close off unneeded rooms
- stuff towels or rags in cracks under doors
- cover windows at night.

Eat and drink. Food provides the body with energy for producing its own heat. Keep the body replenished with fluids to prevent dehydration.

Wear layers of loose-fitting, lightweight, warm clothing. Remove layers to avoid overheating, perspiration, and subsequent chill.

Figure 5.10 US advice: when caught out in a winter storm

Case study: January ice storm, Montreal

In early January 1998, an intense storm brought heavy rainfall to north-eastern USA and south-eastern Canada. When the rain fell in Canada, the ground was freezing. This resulted in five days of freezing rain.

The ice built up on power lines leading, incredibly, to the collapse of 600 giant transmission towers and tens of thousands of pylons. The lines littered the ground and vast areas lost power – an estimated 3 million customers had their power cut off after the storm. With temperatures as low as minus 20°C and no electricity for heating or cooking, the effects of the ice storm were far-reaching. The cost of the damage was put at $350 million – a record for Canada.

Many people died from hypothermia and the inhalation of fumes (mostly carbon monoxide) as people desperately tried to heat their homes using all manner of methods, including portable barbecues. The city of Montreal ground to a halt as the central area was blacked out. Thousands of people were evacuated to schools and gyms where generators provided light and heat. It took weeks before power supplies were restored.

Review

21 List the different types of severe winter weather. Which of these do you think is **a** the most frequent and **b** the most hazardous? Justify your choices.

22 Explain what is meant by the **wind-chill factor**.

23 **a** What were the reasons for the heavy snowfall in New York State in January 1996?

b Would the hazardous situation have ended once the snowstorm had passed?

24 With reference to the Montreal ice storm:
a describe the formation of freezing rain
b identify its hazardous outcomes
c suggest some of the problems resulting from Montreal being effectively shut down.

WEB SEARCH

There are several web sites worth visiting. For information on the ice storm of 1998, see: www.icestorm98.com/

The National Oceanic and Atmospheric Administration (NOAA) at: www. noaa.gov/ has links to many reports on winter storms.

Also, look up the National Snow and Ice Center at 'www.nsic.gov and the Federal Emergency Management Agency at: www.fema.gov/

SECTION D

Extended enquiry: Tracking Hurricane Fran

Hurricane Fran hit the coast of North Carolina on 5 September 1996. The strong winds, torrential rainfall and storm surge led to 37 people losing their lives. Large areas of the coast were devastated by the hurricane. The cost of the damage made it the third most expensive hurricane in US history.

In this enquiry you will plot the course of the hurricane to see the route it took and to appreciate some of the difficulties scientists face when attempting to forecast where hurricanes will make landfall and do most damage.

In compiling your report on Hurricane Fran you should address the following aspects:

■ The track taken by the hurricane.
■ The effects of the hurricane when it made landfall.
■ The difficulty of predicting that landfall location.

To help you complete your report, work through the questions and activities on page 71.

Day and time	Latitude	Longitude	Pressure	Av. wind speed (km/h)
Aug				
29/0000	16.4	53.7	987	105
0600	17.0	55.0	987	105
1200	17.8	56.3	988	105
1800	18.6	57.5	988	105
30/0000	19.1	58.5	991	105
0600	19.4	59.4	991	105
1200	19.8	60.1	989	105
1800	20.2	60.6	990	95
31/0000	20.5	60.9	988	95
0600	20.8	61.2	987	95
1200	21.1	61.4	984	105
1800	21.5	61.7	983	105
Sep				
01/0000	21.7	62.1	978	105
0600	21.9	62.6	982	105
1200	22.2	63.2	982	112
1800	22.5	63.9	981	120
02/0000	22.9	64.7	978	120
0600	23.3	65.7	976	120
1200	23.6	66.7	976	120
1800	23.9	67.9	976	120
03/0000	24.2	69.0	977	120
0600	24.4	70.1	975	127
1200	24.7	71.2	973	127
1800	25.2	72.2	968	135
04/0000	25.7	73.1	961	152
0600	26.4	73.9	953	160
1200	27.0	74.7	956	168
1800	27.7	75.5	952	168
05/0000	28.6	76.1	946	168
0600	29.8	76.7	952	168
1200	31.0	77.2	954	160
1800	32.3	77.8	952	160
06/0000	33.7	78.0	954	160
0600	35.2	78.7	970	105
1200	36.7	79.0	985	65
1800	38.0	79.4	995	35

Figure 5.11 The progress of Hurricane Fran

Figure 5.12 Base map for plotting the track of Hurricane Fran

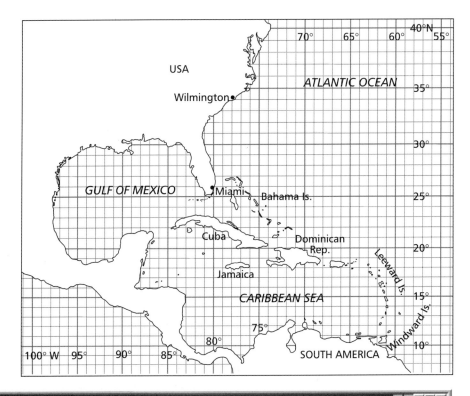

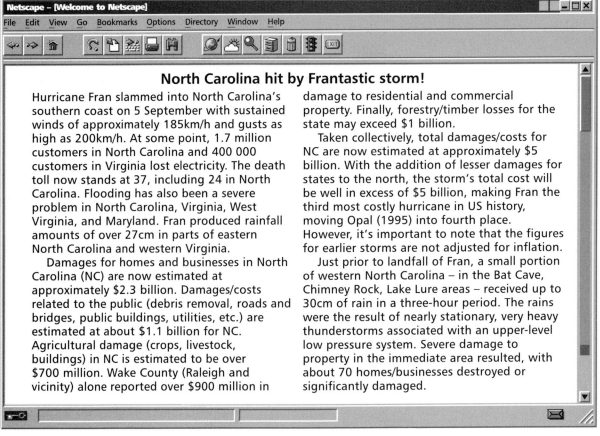

North Carolina hit by Frantastic storm!

Hurricane Fran slammed into North Carolina's southern coast on 5 September with sustained winds of approximately 185km/h and gusts as high as 200km/h. At some point, 1.7 million customers in North Carolina and 400 000 customers in Virginia lost electricity. The death toll now stands at 37, including 24 in North Carolina. Flooding has also been a severe problem in North Carolina, Virginia, West Virginia, and Maryland. Fran produced rainfall amounts of over 27cm in parts of eastern North Carolina and western Virginia.

Damages for homes and businesses in North Carolina (NC) are now estimated at approximately $2.3 billion. Damages/costs related to the public (debris removal, roads and bridges, public buildings, utilities, etc.) are estimated at about $1.1 billion for NC. Agricultural damage (crops, livestock, buildings) in NC is estimated to be over $700 million. Wake County (Raleigh and vicinity) alone reported over $900 million in damage to residential and commercial property. Finally, forestry/timber losses for the state may exceed $1 billion.

Taken collectively, total damages/costs for NC are now estimated at approximately $5 billion. With the addition of lesser damages for states to the north, the storm's total cost will be well in excess of $5 billion, making Fran the third most costly hurricane in US history, moving Opal (1995) into fourth place. However, it's important to note that the figures for earlier storms are not adjusted for inflation.

Just prior to landfall of Fran, a small portion of western North Carolina – in the Bat Cave, Chimney Rock, Lake Lure areas – received up to 30cm of rain in a three-hour period. The rains were the result of nearly stationary, very heavy thunderstorms associated with an upper-level low pressure system. Severe damage to property in the immediate area resulted, with about 70 homes/businesses destroyed or significantly damaged.

Figure 5.13 Internet news

1 Plot the information in **5.11** on a copy of **5.12**. You might also consider showing the changes in pressure and wind speed during the progress of the hurricane.
 - What were the strongest winds and where was the hurricane when these winds occurred?
 - What happened to the wind speeds when the hurricane hit land? Why did this change occur?

2 ■ How many people were killed by the hurricane?
 - What effect did the hurricane have on public utilities?
 - What other damage was done?

3 Tracking a hurricane enables scientists to issue warnings to countries or regions that seem to be in its path. Forecasting is, however, not as easy as it sounds, as hurricanes tend to deviate from their projected routes. Look back at **5.3** to see the different levels of Advisory that can be issued.
 - At 1800 on 3 September a hurricane warning was issued for north-western Bahamas. Locate this time on your tracking map. Project the track from this point using a broken line to indicate why the warning was issued. Describe what actually happened to the hurricane.
 - What negative effects might result from a warning being issued that then turns out to be inaccurate?
 - Computer models forecasting the landfall location in the USA consistently predicted a more southerly route than the hurricane actually took. What evidence is there from your tracking map to suggest that the computer models made reasonable predictions?

WEB SEARCH

Reports on hurricanes including tracking co-ordinates can be obtained on the Internet by contacting the National Hurricane Center at: www.nhc.noaa.gov/

This extended enquiry can be readily adapted for a more recent hurricane.

Flooding

The nature of the flood hazard

A flood is usually defined as a body of water that rises to overflow land not normally submerged. Flooding is a common and generally natural occurrence, but it frequently presents a major hazard to people. Apart from injury and loss of life, flooding may damage or destroy buildings and inundate agricultural land, ruining crops and drowning livestock. Transport systems such as roads and railways may be disrupted and services such as gas, water and electricity may be cut off. Sewerage systems are often severed, posing serious health hazards. In addition, one of the most ironic outcomes of a flood is that there is often a lack of safe water to drink (**6.1**). Other less dramatic but equally important results of flooding include the loss of working hours, general inconvenience, and a considerable amount of worry, anxiety and stress.

Figure 6.1 Women wading through floodwater to collect fresh drinking water

The causes of flooding

Flooding is most commonly associated with rivers overtopping their banks. However, flooding can also occur during and after torrential rainfall, particularly in urban areas where drains may become blocked or storm sewers overwhelmed by the sheer volume of water. Coastal areas are also at risk from flooding, particularly by storm surges and tsunami.

River flooding

A river flood occurs when a river channel is no longer able to contain the water flowing within it. Excess water sweeps over the banks and onto the adjacent floodplain. River flooding is a perfectly natural event and is to be expected on a frequent and fairly regular basis. By building on floodplains and using them for agriculture, it is people who are turning a natural process into a potential hazard.

A range of factors can increase the likelihood of river flooding.

- Heavy and prolonged rainfall which cannot be readily accommodated by a river channel.
- Drainage basin characteristics such as steep slopes, impermeable bedrock, saturated or frozen soil, and deforestation all encourage the rapid transfer of water to a river channel.
- The silting of channels, particularly by excessive soil erosion.
- The blocking of bridges and culvert entrances by debris transported by the river.
- The action of landslips in displacing water in lakes or blocking channels.
- An increase in storm flow due to urbanisation.
- The failure of flood defence structures such as reservoir dams and embankments.
- A reduction in the size of the floodplain (the natural flood storage zone) by development.
- A reduction in channel size.

Most often, it is a combination of several of these factors that results in river flooding.

Torrential rainfall

Very heavy rainfall, often produced by convectional thunderstorms, tropical cyclones or monsoons, can lead to flash flooding. This may also involve river flooding but it is essentially the result of the inability of a drainage system, whether natural or unnatural, to cope with huge amounts of water. In April 1998, the South Midlands in the UK suffered some of the most serious flash floods in living memory (**6.2**).

Figure 6.2 Flash floods in central England

Thousands flee homes as flooding causes chaos

by Will Bennett

Torrential rain at the start of the Easter weekend caused at least £15 million damage across much of central England.

Throughout Thursday night and early yesterday the storm, which claimed up to four lives, caused rivers to burst their banks, halted road and rail communications and caused people to flee their homes. The damage was spread across counties from Herefordshire to Norfolk, with Oxfordshire, Worcestershire and Warwickshire suffering the worst effects of the storm.

RAF helicopters were called out to rescue motorists, campers and fishermen who found themselves trapped by the rapidly rising floodwater. In Pershore, Worcs, John Barnham, a 55-year-old farmer, was winched to safety by helicopter from the roof of his Land Rover after it was marooned while crossing a bridge.

Sonia Broughton, who works at the Abbots Salford caravan park near Evesham, Worcs, said she and her husband, Cyril, 'thought we were about to die' before they were rescued by a helicopter. She said: 'I clung on to a mobile chalet with our dog but an elderly lady could not hold on and was starting to get swept away. My husband and I grabbed hold of her and pulled her back. We managed to get into the chalet but the water was rising all the time.'

Another helicopter was sent to help 19 night anglers who were trapped when they were surrounded by floodwater which rose by 2.75m an hour near Newport Pagnell, Bucks. A fire service boat was launched to rescue some of the fishermen who dived into the water at Little Linford lake to try to swim to the shore.

Floodwater completely cut off Ledbury, Herefordshire, for a time and hotels were soon fully booked by stranded motorists. In Southam, Warwicks, 1000 people were moved from their homes and spent the night in a local school. In Banbury, Oxon, more than 200 people were put up in temporary shelters overnight after being evacuated from their flooded homes. John Batchelor, 54, said: 'I awoke to people shouting and hammering on my front door. I could not believe what was happening. Water was gushing through my entire flat.'

Source: *The Daily Telegraph*, 11 April 1998

Coastal flooding

Low-lying coastal areas are particularly vulnerable to flooding. In addition to rivers bursting their banks, coastal areas bear the brunt of the heavy rainfall and storm surges associated with tropical cyclones (see page 57). Whilst such events are commonly associated with countries like Bangladesh, storm surges do occasionally affect the coast of the UK. In 1953 a deep area of low pressure tracked south through the North Sea causing flooding along much of the coast of eastern England, particularly in East Anglia. Over 300 people were killed by the floods and 35 000 people had to be evacuated.

Tsunami, huge waves associated with earthquakes and volcanic eruptions, can cause tremendous damage and loss of life in coastal areas. Tsunami are generated by a sudden displacement of the ocean floor which sends waves out in all directions. As the waves approach the shore, friction with the shallowing sea floor causes them to assume great heights. Water is often drawn back from the coast in an alarming fashion. The first wave is usually the most destructive, but waves can continue to batter a coastline for up to eight hours after the earthquake or eruption. The force of the moving water is enormous and can cause tremendous destruction.

Some twenty countries around the rim of the Pacific Ocean have been regularly affected by tsunami and an estimated 50 000 people have been killed by them this century. Japan has been particularly badly affected. In 1933 a giant 24-metre wave killed 3000 people and destroyed 5000 houses.

Review

1 Identify some immediate and some longer-term effects of flooding.

2 Contrast the effects of flooding in developed and developing countries.

3 In what ways can the following drainage basin factors increase the likelihood of river flooding:
   ■ steep slopes
   ■ frozen soil
   ■ impermeable bedrock
   ■ deforestation?

4 Outline the causes of coastal flooding.

5 'Of all the human-influenced causes of flooding, it is urbanisation that has the most significant effect.' Comment on the validity of this statement.

Case study: River Mississippi (USA), 1993

In the summer of 1993, the fifth largest drainage system in the world was the scene of one of the worst natural disasters in the history of the USA. Some 40 per cent of the country is drained by the Mississippi and its tributaries. In total 48 people lost their lives, over 50 000 people had to leave their homes, and an estimated $11 billion worth of damage was caused (**6.3**). The effects of the flooding were felt nationwide as airports and bridges were closed, trains had to be re-routed and some 2000 barges were left stranded along a 900km stretch of the river. Seven million hectares of some of the most productive land in the world were affected, leading to massive losses of food crops, and inflated market prices.

State	Flood-related deaths	Property damage (US$ millions)	Agricultural loss (US$ millions)
Minnesota	4	51	865
North Dakota	2	100	420
South Dakota	3	25	725
Iowa	5	1250	450
Nebraska	2	50	292
Kansas	1	160	434
Wisconsin	2	101	800
Illinois	4	930	565
Missouri	25	2000	2000
Totals	**48**	**4667**	**6551**

Figure 6.3 Flood damage statistics, 1993

The cause of the flooding was a combination of natural and unnatural factors. There was a tremendous amount of rainfall throughout June and July. The heavy rain followed a wet autumn and a lot of heavy snow during the winter. The soil was already saturated and was unable to soak up all the additional rainfall. River levels rose alarmingly in the upper reaches of the Mississippi and the Missouri (**6.4**). At St Louis, flood crests reached 24 metres above their average height.

For much of its length, the Mississippi can hardly be thought of as a 'natural' river. Throughout the early part of the 20th century, the US Army Corps of Engineers constructed levées, storm walls, wing dykes and artificial banks with the intention of taming the river to allow for extensive floodplain development. Whilst these measures coped admirably with small-scale floods, they made for very hazardous situations when higher discharges occurred such as those experienced in 1993. The artificial river expanded to such a size that when it did eventually burst its banks, it did so in a far more devastating fashion than if it had been allowed to do so earlier and further up the drainage basin system.

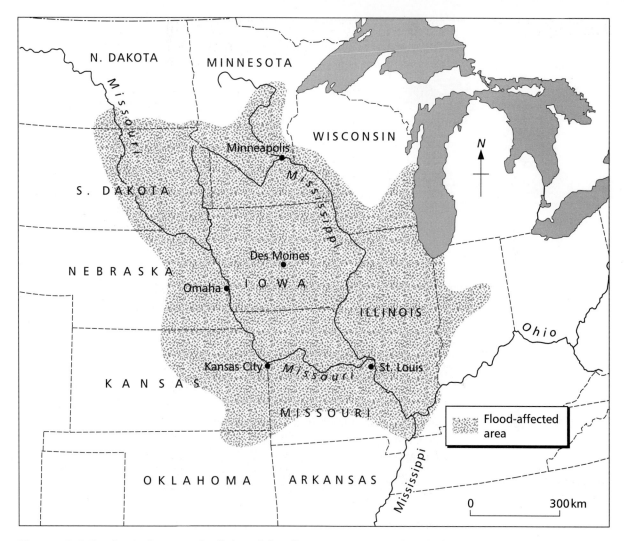

Figure 6.4 Flooding in the upper Mississippi and Missouri valleys

In the light of the disaster, some people advocated demolishing the flood control works. They argued that the river should be allowed to flood naturally, thereby preventing the huge build-up of water that caused the devastating flood of 1993. Of course, if the river defences were to be dismantled, those people living next to the river would be under a greater threat of more regular but smaller-scale flooding.

Case study: Imlil (Morocco), 1995

On 17 August 1995, a torrential rainstorm dumped 70mm of rain in just over two hours close to the village of Imlil near Marrakech. The flash flood that followed caused a 27-fold increase in the discharge of the Reraya river (**6.5**). Boulders the size of lorries were swept down the valley by a wall of water some 6m high. Water was diverted through the village where it washed away cars and damaged many buildings. About 150 people were killed by the event which lasted for only three hours before the river returned to its normal level.

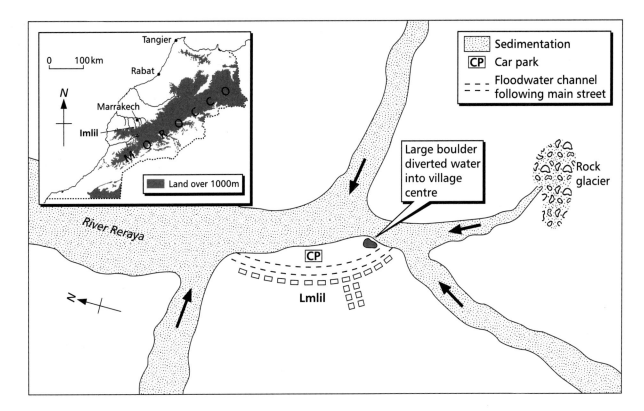

Figure 6.5 The flash flood at Imlil

The flood was exacerbated by deforestation of the valley sides for fuelwood. The lack of trees had left the soil hard and impermeable so that the rainwater flowed extremely rapidly into the rivers.

The flood caused significant long-term damage to the farming sector on which many of the local Berber people depend.

- Fields of maize, a staple food crop, and alfalfa and grass, grown for fodder, were devastated.
- A large amount of topsoil was washed away.
- Many head of goats, mules and cattle were lost. Mules are especially important as they do most of the work in the fields.
- Terraces were destroyed.
- Many walnut trees, which form an important cash crop, were destroyed. It takes 10–15 years for a walnut tree to mature.
- Irrigation channels, essential for cultivating food crops in such an arid region, were damaged or became choked with debris from the flood.

Case study: Papua New Guinea, 1998

On 17 July 1998 the north coast of Papua New Guinea was hit by several tsunami, the greatest of which was a 10-metre-high wall of water. The tsunami, triggered by a magnitude 7 undersea earthquake, tore through remote coastal villages reducing the flimsy houses to matchwood and washing thousands of people into the sea. Witnesses described the 10-metre-high wave as sounding like a jet engine as it raced towards the coast.

The eventual death toll is expected to be more than 3,000, although the area is so remote and undeveloped (most people are subsistence farmers) that an accurate final figure will never be known.

In the days and weeks that followed, disease became a serious problem as bodies began to decompose in the tropical heat. In addition, crocodiles feasting on the dead became a serious threat to rescuers.

News of the seriousness of the disaster took several days to reach the outside world due to poor communications and lack of infrastructure in this part of Papua New Guinea. Clearly, had outside aid been able to reach the stricken area sooner (particularly from nearby Australia), fewer people would have died in the aftermath of the tsunami.

Review

6 a What were the natural causes of the 1993 Mississippi floods?
  b On a sketch map of the area shown in **6.4**, select a suitable technique to show the information in **6.3**.
  c In what ways did human intervention make the floods worse than they would otherwise have been?
  d Suggest the arguments for and against the major engineering methods of 'taming' rivers. Use the Mississippi case study to stimulate your ideas.

7 a Why is the Imlil flood a good example of a flash flood?
  b What human factor contributed to the scale of the flood?
  c Describe some of the longer-term effects of the flood.
  d What would be the most appropriate form of aid that could be offered to the villagers?

8 a Describe the short-term and longer-term effects of the tsunami that hit Papua New Guinea in 1998.
  b How did the poor infrastructure in the region increase the scale of the disaster?

SECTION D

Reducing the flood hazard

River flooding

A recent study has revealed some interesting perceptions of the flood hazard and of management options. A questionnaire survey of 3000 floodplain residents in the UK came up with perhaps a surprising conclusion. People living closest to a river (and therefore subjected to the greatest risk of flooding) were less happy to have a flood defence scheme constructed in their area than people living further away. The main reason for this was that the riverside residents felt that a defence scheme would be likely to damage the quality of their local environment.

Traditionally, however, it has been the so-called 'hard-engineering' flood defence schemes that have been most widely employed by local authorities in their attempts to reduce the risk of flooding. Figure **6.6** illustrates some of the more common structural flood alleviation schemes. Most schemes aim to stop a river flood occurring in an urban area. Whilst this may well be achieved, such schemes often simply shift the problem further downstream. For example, an enlarged channel may speed water past a town only to pond-up and then flood further downstream where the natural channel cannot cope with the sheer volume of water.

Structural scheme	Characteristics	Diagram	Examples
Raised riverbanks	1 One of oldest methods. 2 Often constructed of river sediment. 3 May be concrete lined. 4 Generally have slopes of 1 in 2 with a 2m flat upper surface. 5 Sometimes unpopular with residents due to loss of visual amenity.	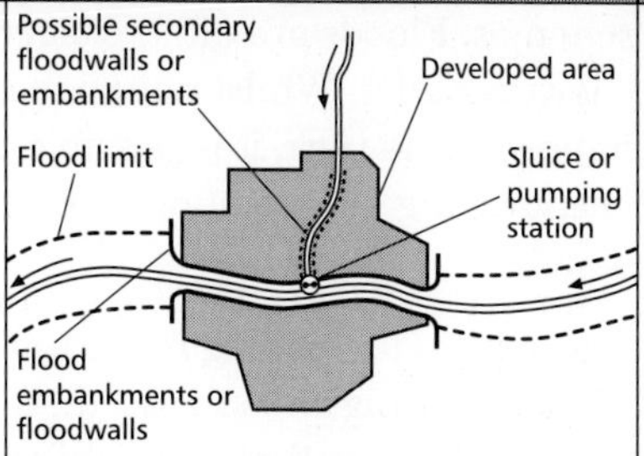	Numerous – most towns with rivers running through them have raised banks.
Channel enlargement	1 Aims to improve the flow of water in the channel thus preventing over-topping. 2 Involves dredging the channel to increase its cross-sectional area. 3 Removal of trees and bushes along the banks can increase the speed of flow, but at what environmental cost?	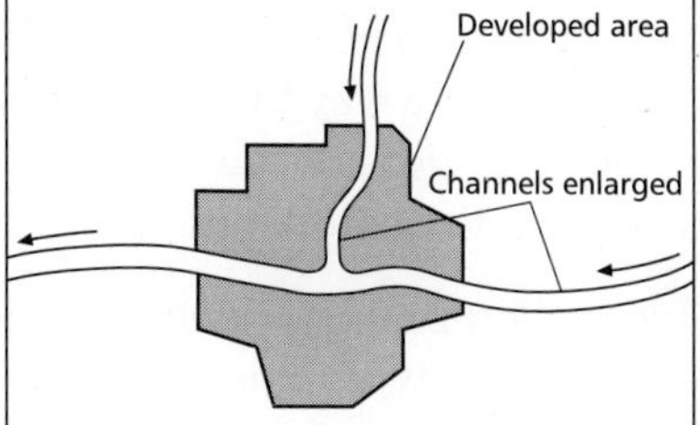	River Tame in Birmingham was enlarged in the 1930s, but it was soon filled in during low flows (the enlarged channel led to increased rates of deposition). By 1959 it had reverted back to its original size!
Flood relief channels	1 Expensive option but can be good if the river stretch running through a town cannot easily be modified. 2 Sluice gates control water entering the relief channel, with normal flows continuing to pass down the original channel.	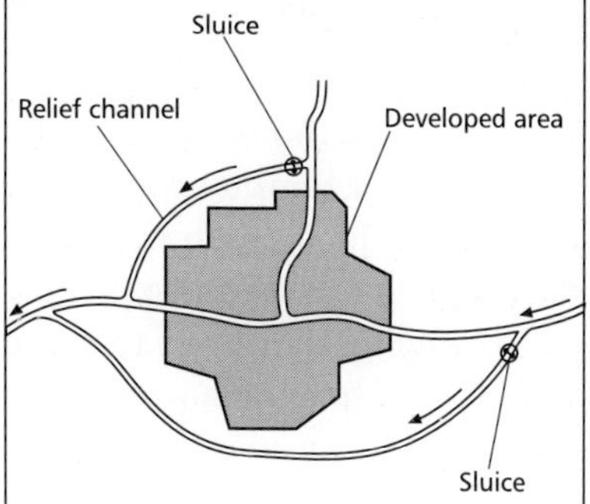	1 Great Ouse, between Downham Market and King's Lynn. 2 Coronation Channel at Spalding (Lincs). 4.8km long and 27m wide, it was constructed to take floodwater from the River Welland around the town. 3 There are plans to construct a major flood relief channel alongside the River Thames between Maidenhead and Windsor.
Catchwater channels	1 Channels designed to reduce flood risk by diverting the entire flow away from the town. 2 Often rejoin the same river at a point downstream.	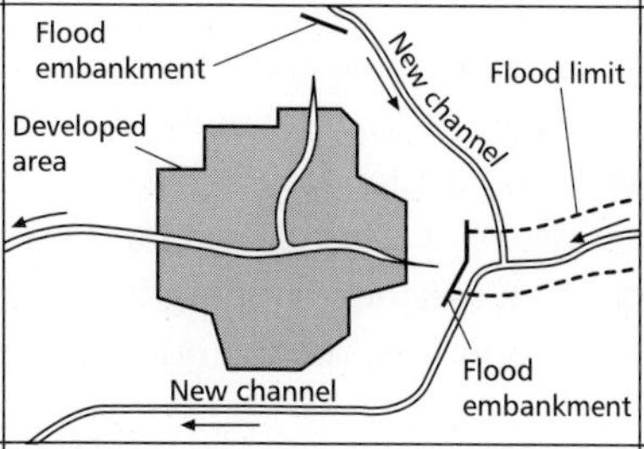	1 Romney Marsh, Kent. 2 River Avill, near Dunster (Somerset).
Flood storage ponds	1 Water stored for a short time to regulate river flow. 2 Can be 'onstream' or 'offstream'. Onstream ponds involve the river flooding above a pond. Offstream ponds take water onto the floodplain into an embanked area.	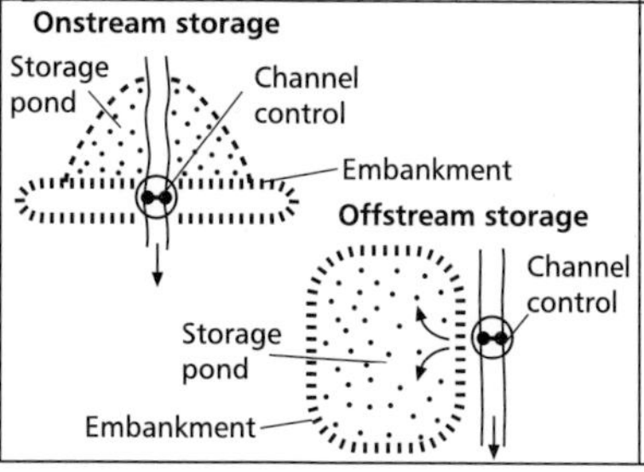	1 River Stort in Hertfordshire. 2 Ashford, Kent.

Figure 6.6 Some structural schemes for flood alleviation

Whilst structural engineering schemes may have their place, there has been an increasing trend towards 'soft-engineering' or 'behavioural' schemes. Floods are often exacerbated by certain drainage basin characteristics. Whilst not a lot can be done about steep slopes or an impermeable rock, it is possible to modify land use, for example by encouraging afforestation or by changing from an arable to a pastoral use of fields.

There is an increasing move towards restoring rivers to their natural state. Whilst flooding will still occur, it will be on a smaller scale. The restoration of floodplains to their original state will increase the interception of storm runoff and will provide natural storage ponds for excess water, thus preventing flood damage elsewhere. The Mississippi flood of 1993 illustrates the way in which channel modification may well increase the flood hazard (see page 76). Although this may make small floods less frequent, the potential for larger, more catastrophic floods is increased.

Floodplain zoning, which is particularly common in the USA, severely restricts the type of development allowed on the floodplain. As a result, the land adjacent to rivers is often used as public open space and car parks rather than sites for housing and industry. This approach to the flood hazard accepts flooding as being a natural process and focuses instead on modifying human behaviour rather than that of the river.

A range of basin instruments (both hydrological and meteorological), satellites and computer modelling can be used to warn people of an impending flood. Such warnings can be used to trigger management decisions (for example to open sluice gates to release water into a storage pond) which themselves may reduce the flood risk. Alternatively, warnings can be used to enable people to protect property using sandbags or, if necessary, to evacuate an area completely. In the UK the Environment Agency has three levels of warning which are relayed to the police and to the public:

- Yellow (flooding is possible) – rivers are running high and flooding is likely, especially of roads and farmland.
- Amber (flooding is likely) – flooding of a number of roads, considerable areas of agricultural land and some high-risk properties is likely.
- Red (serious flooding is likely) – flooding of a significant number of properties, roads and large areas of agricultural land is likely.

Case study: Whitland flood alleviation scheme (Wales)

The small town of Whitland lies roughly midway between Haverfordwest and Carmarthen in the county of Dyfed in south-west Wales. The town grew up at the bridging points of the River Taf to the south and the River Gronw to the east (**6.7**). However, as it developed on the adjacent floodplains of the two rivers, so it became more vulnerable to flooding.

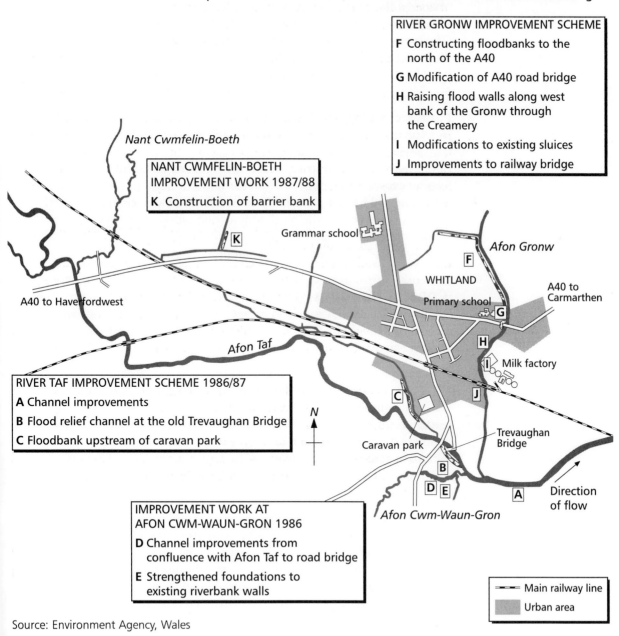

RIVER GRONW IMPROVEMENT SCHEME

F Constructing floodbanks to the north of the A40

G Modification of A40 road bridge

H Raising flood walls along west bank of the Gronw through the Creamery

I Modifications to existing sluices

J Improvements to railway bridge

NANT CWMFELIN-BOETH IMPROVEMENT WORK 1987/88

K Construction of barrier bank

RIVER TAF IMPROVEMENT SCHEME 1986/87

A Channel improvements

B Flood relief channel at the old Trevaughan Bridge

C Floodbank upstream of caravan park

IMPROVEMENT WORK AT AFON CWM-WAUN-GRON 1986

D Channel improvements from confluence with Afon Taf to road bridge

E Strengthened foundations to existing riverbank walls

Source: Environment Agency, Wales

Figure 6.7 Whitland flood alleviation scheme

In recent years, floods have been common in Whitland (1979, 1981, 1986 and 1991). The event of August 1986 was one of the most serious. About 100mm of rain fell on the River Gronw basin in a single day due to the tail-end of Hurricane Charley. The Gronw flooded just north of the A40 road bridge and water flowed across the school playing fields before engulfing the town (**6.7**). Some 220 properties were inundated including houses and industrial buildings. Traffic was disrupted and vast areas of agricultural land were flooded. Damage was put at £1.4 million.

The 1986 flood occurred after some improvements had already been made to the river. Clearly, these improvements were inadequate and as a result, a revised and extended scheme was designed and implemented. The general principle is to prevent water escaping from the rivers. This has involved the following works:

- bypassing existing obstructions to flows
- enlarging existing bridge openings and improving their entry and exit conditions
- constructing walls and embankments to contain floodwaters.

Work was completed in 1993 at a total cost of £512 000, and the town is now protected from a 1 : 100-year flood (a flood of a magnitude such that it would have a probability of occurring only once in a hundred years). In addition to the hard-engineering works, there has been a programme of tree planting and bankside landscaping, while a nature area has been set aside for the local primary school.

Coastal flooding

Reducing the flood hazard associated with storm surges has already been considered (page 58). Similar measures are taken to reduce the hazard posed by tsunami (page 23).

Forecasting and public warnings are employed widely across the world. There are a number of tsunami warning centres, for example the Pacific Tsunami Warning Center located near Honolulu, Hawaii. This monitors earthquakes in the Pacific basin and, with the aid of tracking methods and mathematical models, attempts to predict when and where a tsunami will hit. An approximation of the wave height can also be calculated. Warnings can then be issued so that people at risk can be evacuated to high ground.

Many coastal areas at risk from tsunami have emergency plans, involving local warning systems and evacuation procedures. The basic advice given is fairly obvious – stay away from low ground until the threat has subsided. Boats out to sea are advised to stay there, as it is only close to shore that the tsunami wave builds to become a major threat. Harbours are particularly at risk, and structures such as sea walls and offshore barriers have been constructed especially in Japan which is very vulnerable to tsunami.

9 a Many riverside residents in the UK favour the 'natural-look' river. However, such a view is not universal. Read the following short extract and suggest why local residents in the Portuguese town of Setubal prefer the option of culverts (underground pipes).

In Setubal ... the concrete-lined river represents an environmental nuisance for a major part of the year. When there is a low flow or no flow at all, sewage contamination becomes obvious and the dumping of rubbish makes the river an eyesore. For the local riverine population there is the additional problem of flooding during times of heavy rainfall.

b Culverts can sometimes get blocked during times of high flow. Why might this increase the risk of flooding?

10 *There is little point in consulting local people about flood defence as they have a limited and highly subjective knowledge and understanding of the issues involved. Water managers should do what is necessary and not worry themselves unduly about the views of local people.*

Do you agree or disagree with this view? Give reasons for your answer.

11 Study **6.6**.
a Explain how each structural scheme reduces the risk of flooding.
b For each scheme try to identify some advantages and disadvantages.

12 Explain how river basin modifications can reduce the flood hazard.

13 a Explain what is meant by 'river restoration'.
b Do you consider it to be a suitable option for rivers such as the Mississippi?

14 *The government of Morocco has, for many years, been committed to a programme of reforestation in areas of serious soil erosion in the High Atlas. As part of this programme, bottled gas is heavily subsidised to discourage the use of wood for fuel by local people. However, fuelwood is still collected by some families either for reasons of poverty, or because of the popular belief amongst the Berber that wood fires improve the taste of the food being cooked.*

a Why is the Moroccan government encouraging reforestation?
b Why is the subsidisation of alternative fuels not working with all families?
c What do you think the government should do to ensure that its reforestation programme is more effective?

15 Study **6.7**.
a Why does the site of the town increase the flood risk?
b What is the main purpose of the barrier bank at M?
c Identify some modifications to the existing river channels that took place as part of the scheme, and suggest reasons for their use.
d In what ways, and for what reasons, do you think that some of the bridges were modified?

Extended enquiry: Protecting Bangladesh from flooding

Bangladesh faces enormous problems relating to flooding, probably more so than any other nation. The majority of its large and rapidly-growing population (currently over 110 million and set to double by the year 2030) live on floodplains and deltas at, or even below, sea-level.

Flooding is both a blessing and a curse. Water is important for the floodplain fisheries (a key source of protein in a country where poverty is widespread) and for the cultivation of rice and other food crops. Floodwaters bring fertile silts to nourish the adjacent fields. However, floods can be devastating, causing massive loss of life and serious long-term effects such as homelessness, disease and unemployment.

There have been some attempts to prevent flooding, but lack of money and the sheer scale of the problem mean that success has been very limited. For many farmers, of course, the absolute prevention of flooding is not desirable as they depend on some flooding for farming. There is often a conflict between the more wealthy city dwellers who want to prevent flooding completely, and the poor rural dwellers who want to retain a regime of seasonal but controlled flooding.

In this extended enquiry you will produce a report that examines some of these issues. You should address the following questions:

- What have been the effects of flooding on the people of Bangladesh?
- Why is flooding such a common and devastating problem in Bangladesh?
- What is the Flood Action Plan and what has it achieved?
- What are the future prospects for the people of Bangladesh?

Take time to read through and scrutinise the following materials (**6.8–6.12**).

Figure 6.8 Some background

Flooding in Bangladesh

Flooding is a major problem in Bangladesh. One of the main reasons is that it is very low-lying – 80 per cent of the country comprises the floodplains of three major rivers: the Ganges, the Brahmaputra and the Meghna. These rivers have their headwaters outside the country and many streams have their sources high up in the Himalayas (**6.9**). Huge volumes of water surge into Bangladesh after storms or following snowmelt, causing floods.

Bangladesh is often affected by tropical cyclones. In recent years, several cyclones have had devastating effects. In 1987, 1657 people were killed and in the following year a further 2379 lost their lives. Cyclones in 1991 killed an estimated 140 000 people and in 1997, over 2 million people were affected by a cyclone which hit the Chittagong region (see page 60). The cyclones advance from the south and tend to inflict most damage on coastal regions.

During the monsoon season (June to September), heavy rainstorms often bring widespread flooding. Ironically, this may result from water ponding-up against embankments unable to flow into rivers. Heavy rains in the mountains may, however, cause rivers in Bangladesh to burst their banks.

In addition to the climate and the physical geography of the country, there are other factors encouraging flooding. Widespread deforestation in the headwater regions of the major rivers has increased the rate of soil erosion. This in turn has increased the rate of silting in the rivers, so reducing channel size and encouraging flooding. Poor maintenance of embankments over the years has meant that they have been unable to withstand high flows and they have often given way. Embankments have also been deliberately breached by farmers wishing to make use of water for farming or fisheries. Once weakened, the embankments are less likely to resist high flows in the future.

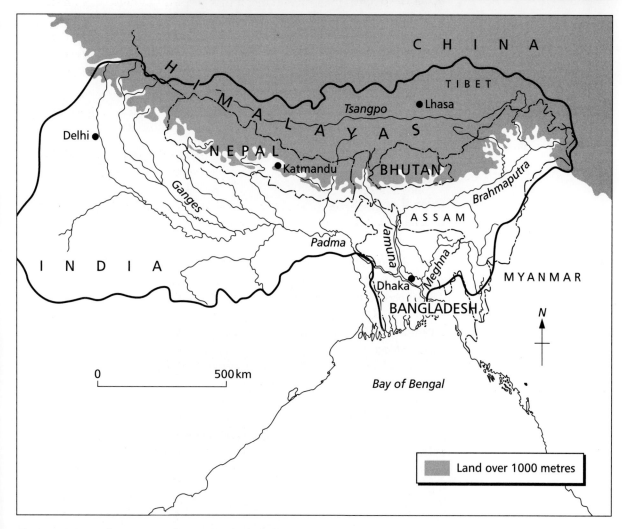

Figure 6.9 Catchment area of rivers in Bangladesh

Figure 6.10 The impacts
of the 1987 and 1988 floods

	1987	1988
Areas flooded (km²)	57 000	84 059
Deaths	1657	2379
Rice production lost (million tonnes)	3.5	2.0
Crop damage (million ha)	no data	7.16
Head of cattle lost	64 700	172 000
Poultry lost	206 000	410 000
Houses destroyed or damaged (millions)	2.5	7.2
Hospitals flooded	no data	45
Clinics flooded	no data	1400
Schools flooded	no data	8481
Industrial units damaged	no data	14 000
Trunk roads damaged (km)	1523	2935
Rural roads damaged (km)	15 107	65 892
Road bridges damaged	1102	898
Railway lines damaged (km)	no data	638
Railway bridges damaged	no data	34
Levées damaged (km)	1279	1990
Overall cost US$m	**1211.7**	**1138.1**

Figure 6.11 Past flood
protection measures

*About 6000km of flood embankments already existed pre-1990. Both sides of the Ganges river are fully embanked. Most of the Brahmaputra right bank and about half of the left bank are also embanked, as well as substantial sections of the Padma (the combined Brahmaputra and Ganges) and Meghna rivers (see **6.9**). So are many eastern rivers which are subject to flash floods from adjoining hill areas. In addition, several thousand kilometres of embankments exist around polders in the coastal zone. A number of inland areas have been also empoldered, some of them provided with pump drainage.*

While there have been some successes, such as the Chandpur Irrigation Project, there have been many partial or total project failures. The Flood Action Plan (FAP) 12/13 studies found that, amongst many contributory factors, the main reasons for poor performance were:

- *inadequate attention to internal drainage behind embankments (flooding results from heavy local rainfall as well as river floods)*
- *inadequate fund allocations for maintaining structures*

- *the blue-printing of projects over the heads of local residents*
- *conflicts between farmers and fishermen, and between those living outside and those inside embankments.*

The consequence has often been public cutting of embankments during floods, with substantial loss of benefits and even additional damage.

Attempts are being made under the FAP to strengthen consultation and public participation. This is not an easy task in a strongly hierarchical society, and it would be unrealistic to expect overnight success.

Source: Hugh Brammer in *Tiempo* 8, April 1993, University of East Anglia

Figure 6.12 The Flood Action Plan

The Flood Action Plan (FAP) is co-ordinated by the World Bank, at the request of the Government of Bangladesh. The cost of the first phase (1990–95) is about US$150 million. Local co-ordination is through the Flood Plan Co-ordination Organisation (FPCO) set up under the Ministry of Irrigation, Water Development and Flood Control.

The aims of the first five-year phase are:

- *to establish the principles and criteria for sustainable flood mitigation*
- *to undertake comprehensive planning studies, and*
- *to begin the implementation of high-priority projects.*

Contrary to assumptions made by some of its critics, the FAP is not a construction plan. It mainly comprises studies and pilot projects which aim to identify the most appropriate flood mitigation measures, non-structural as well as structural, for different parts of the country. The objective is to identify projects for donor funding. Non-structural activities include the strengthening of existing flood warning, disaster management and hydrological modelling systems. Reviews have been made of the past performance of flood protection projects and of how people responded to the 1988 floods.

A number of special environmental studies have been made, and a comprehensive study of floodplain fisheries is in progress. Guidelines on project economic assessment, environmental impact assessment and public participation have been prepared for FAP consultants. Projects have been assisted in obtaining topographical surveys, aerial photography and satellite imagery, and a geographic information system has been set up.

Structure-linked activities include feasibility and design studies for the rehabilitation of two major embankments. These are the Brahmaputra right embankment, frequently breached by river erosion, and the south-eastern section of the coastal embankment which had fallen into disrepair (and was subsequently virtually destroyed by the 1991 tropical cyclone before rehabilitation had started). Feasibility studies have also been carried out for works to protect Dhaka city against floods and several regional towns (including Chandpur) mainly against river erosion.

To date, three out of five regional planning studies have been completed. These have led to feasibility studies being made in three identified flood control project areas which are expected to proceed to detailed design in 1993. Ultimately, it is envisaged that the regional flood management plans will be integrated into the National Water Plan.

The FAP does not specifically address possible impacts of global warming. However, since all climate models forecast increased monsoon rainfall for the Bangladesh region as global warming develops, floods may become even more frequent and severe than at present in future decades.

Except for urban areas, the FAP does not advocate total flood protection. For rural areas, the policy is that of 'controlled flooding'. River embankments will be provided with regulators to allow the continuation of the 'normal' flooding to which farmers are accustomed and which provides benefits to fisheries, soil fertility and navigation. The regulators would be closed to keep out unwanted high or untimely floods which cause damage. Secondary embankments behind main embankments would divide the protected area into compartments, enabling water flow across the land to be controlled. The objective is to give farmers a more secure environment for investment in crop production in the monsoon season.

One of the principles of environmental management adopted is that any people who might suffer adverse effects as a result of a project intervention should be adequately compensated.

An example being studied is that of people living on unstable alluvial islands (chars) in the main rivers. There is concern that such people might suffer from increased flood levels and channel instability if new embankments confine the main channel. It is impossible to resettle such people elsewhere. There is nowhere for them to go. (Ideally, they should not have been allowed to occupy such hazardous land in the first place.) One form of 'compensation' being considered is flood proofing: for example, raising the level of homestead mounds, providing flood refuges, and strengthening community services.

The single most expensive item (about US$40 million) in the FAP programme is the river training pilot project (FAP21/22). This aims to test methods of river training and bank defence on selected reaches of the Brahmaputra river. The active channel of this river is 10–15km wide. It is strongly braided, constantly creating new chars and washing other land away.

Historically, the Brahmaputra has shown a preferred tendency to erode its right bank, advancing westward at an average rate of about 100m a year. This erosion has destroyed about half of the 220km length of the Brahmaputra right embankment built in the 1960s, necessitating recurrent rebuilding of threatened or eroded sections. Encroaching onto long-settled farmland, this erosion causes great distress to established *families who are rendered landless. Understandably, there is strong public pressure to halt this erosion.*

This is not an easy task. The Brahmaputra is a formidable river. So are the Ganges, Meghna and other rivers which also suffer bank erosion. The rivers run through soft alluvial sediments, and can scour channels 30–50m deep. There is no suitable rock for making structures. The cost of building groynes and revetments is, therefore, very high.

One approach to be tested is that of selective intervention (such as closing off or dredging channels) to 'steer' the river away from critical bank sections. The Chinese appear to have had some success with this technique on the Hwang Ho river.

Will the FAP succeed? It is too early yet to be certain, one way or the other. However, whether additional embankments are eventually built or not, considerable benefits should accrue from the greater technical knowledge acquired from the multidisciplinary studies and from new approaches in public participation, environmental impact assessment, flood proofing, and so on.

One has to ask, what is the alternative to flood protection in Bangladesh? The population, now over 110 million, is expected to double by around 2030. That population lives predominantly on the country's floodplains which occupy 80 per cent of the total area. Agriculture is the mainstay of the economy and seems likely to remain so.

Source: Hugh Brammer in *Tiempo 8*, April 1993, University of East Anglia

Enquiry

1 ■ Describe the natural causes of flooding in Bangladesh (the climate).
■ Look back to Chapter 5, Section A to find out about the cyclone threat to Bangladesh.
■ What human factors may contribute to the flood hazard?
■ In what ways does flooding bring benefits to farmers?
■ Identify the short-term and longer-term effects of cyclones.
■ In what ways does the fact that Bangladesh is one of the poorest countries in the world exacerbate the problems?

2 ■ What measures of flood protection have been employed in the past?
■ Suggest reasons for the measures adopted.
■ Why have they not always been successful?
■ How important is it to involve local people in the decision-making process?

3 ■ What were the aims of the Flood Action Plan?
  ■ Describe the technology that was used in the research.
  ■ Give some examples of the types of study that were carried out.
  ■ Describe the plight of the people who live on *chars*. Why might channel improvements increase the risk of flood damage to these areas?
  ■ What is meant by 'controlled flooding'? How does this measure of protection accommodate the needs of the local people?
  ■ What is meant by 'river training' and what does it involve?

4 ■ What should be done to advance the work of the Flood Action Plan?
  ■ Should Bangladesh be totally protected from flooding?
  ■ How might global warming and sea-level rise increase the flood hazard in the future?

5 Conduct a Web search to find out the latest information about flooding and flood protection in Bangladesh. You might also investigate some of the many global warming sites.

Conclusions

Converting natural events into disasters

Natural hazards involve the interaction of naturally occurring physical processes and human activity. It is important to accept that the processes themselves are entirely natural. For example, slopes achieve a state of equilibrium by processes such as mass movement, and rivers naturally respond to high volumes of water by overtopping their banks. Earthquakes and volcanic activity have been occurring since the formation of our planet and they have played a major part in determining many of its features. Atmospheric phenomena such as winter storms, tropical cyclones and tornadoes are also entirely natural events.

These natural events do, however, pose threats to people. They represent potential hazards. As the population of the world has grown, so the hazard risk and the risk of disaster resulting from natural events have increased. Some of the largest cities in the world are now at risk from tropical cyclones, tsunami and earthquakes. One of the greatest challenges to people today is learning to live with hazards and trying to minimise the risks they pose.

Nearly a third of all natural disasters involve flooding. Many of the hazards studied in this book give rise to flooding, for example tropical cyclones, torrential rainstorms and tsunami. Flooding will probably continue to be the major natural hazard in the future, particularly as so many people live on or close to the coast or alongside rivers where the threat of disaster is greatest. About 20 per cent of all disasters are related to tropical cyclones. These events are devastating, particularly for poor countries such as Bangladesh, which not only suffer massive injury and loss of life but are least well equipped to cope with the aftermath of such a disaster.

Counting the costs

The long-term effects of disasters – such as homelessness, shortages of water and food, loss of livelihood, together with stress and anxiety – should not be underestimated. The media, quite understandably, tend to focus on the immediacy of particular events. After two or three days, the attention fades as new stories emerge, and the long-term plight of the homeless, the unemployed and the emotionally distressed fails to reach the eyes or ears of the world. Perhaps one reason for this is the inability to quantify the long-term effects of a natural disaster. It is relatively easy to count dead bodies but far less easy to measure the true effects of homelessness, unemployment and family bereavement on those who have survived a natural disaster.

Two worlds

Natural disasters affect rich and poor countries alike but the effects can be dramatically different. When an earthquake, tornado or storm surge hits a country like the USA, the death toll is amazingly light yet the cost of damage in monetary terms is extremely high. When a similar event hits a poor country like Bangladesh, there may be less monetary damage but often a far greater human tragedy unfolds. For example, in 1993 and 1994 two earthquakes of similar magnitude hit India and the USA respectively. The Los Angeles earthquake resulted in only 40 dead whereas the Indian earthquake killed 25 000 people. Tornadoes in Texas killed 27 people in 1997. A year earlier, an estimated 700 people in Bangladesh lost their lives during a tornado.

Clearly some people are at greater risk than others from natural disasters, particularly if they live in developing countries. In developed countries, people have the luxury of quite sophisticated warning systems and massive emergency aid programmes if disaster should strike. Extensive research and the implementation of mitigation methods help to reduce the impact of natural hazards in these richer countries. In poorer countries, however, there is a basic lack of money and technology both to prevent the damage done by natural events and to cope with the aftermath. Research has shown that the four highest-risk nations based on past disasters are Bangladesh, Nicaragua, Guatemala, and Honduras. All four are low-income countries.

The human hand in hazards

Whilst natural events often have significant effects on people's lives, in some cases people can have significant effects on the natural processes themselves. For example, many of the disasters associated with landslides have had a human component in their cause. The Aberfan landslide of 1966 was largely the result of coal waste being inappropriately dumped on top of a spring. The Vaiont landslide of 1963 resulted from the inappropriate choice of a site for a dam and reservoir. Construction on slopes the world over has led to devastating landslides as slope angles have been altered, drainage impeded and extra weight added, making slopes unstable.

Human activity has also been linked with triggering earthquakes. Mining, water abstraction and reservoir construction have all given rise to increased earthquake activity. There is even some evidence to suggest that the devastating Indian earthquake of 1993 may have been caused, at least in part, by nearby reservoir construction.

Many incidents of flooding have been linked to human activity. Deforestation is known to speed up the transfer of water in a drainage basin and it can also lead to increased rates of soil erosion which may in turn clog river channels. River channel modification, such as that which took place on the Mississippi river, may have prevented the frequent minor floods that plague riverside settlements. However, in preventing the safety-valve effect of minor flooding, high-magnitude events have been encouraged. Had the channels been left alone, small-scale flooding would have prevented the build-up of water that led to the disastrous floods in 1993.

The way ahead

Natural events will continue to become hazards and to create disasters. As the population of the world increases, so the risks and threats posed by natural hazards will increase also. There is evidence too that some of the natural processes may be intensifying due to global warming, although it is far too early to judge for certain. Instead of trying to control the natural processes, we should learn to live with them, being prepared to modify and adjust our attitudes and behaviour. We are a part of nature and the global systems. We should not have the arrogance or naivety to assume that we can control them. We need to learn to understand and respect natural processes so that, whilst they may remain potential hazards, the risks to people and property are minimised.

Further reading

There are many books on natural hazards. The following is not intended to be an exhaustive list but rather a selection of the variety of books available and the different approaches adopted in studying the topic.

Frampton, S. (1996) *Natural Hazards*, Hodder & Stoughton

McCaulay, J. (1987) *Nature Fights Back*, Longman Paul

Manuel, M. *et al.* (1997) *Hazards*, CUP

Morgan, M. (1997) *Weather and People*, Prentice Hall

Robinson, A. (1993) *Earthshock*, Thames & Hudson

Smith, K. (1996) *Environmental Hazards*, Routledge

Waltham, A.C. (1994) *Foundations of Engineering Geology*, Blackie Academic

Whittow, J. (1980) *Disasters*, Pelican

Several good articles have been published in the *Geography Review* and in *Geofile*. The World Wide Web is an outstanding source of information on natural hazards.